Update on Mesenchymal and Induced Pluripotent Stem Cells

Edited by Khalid Ahmed Al-Anazi

Published in London, United Kingdom

IntechOpen

Supporting open minds since 2005

Update on Mesenchymal and Induced Pluripotent Stem Cells
http://dx.doi.org/10.5772/intechopen.77857
Edited by Khalid Ahmed Al-Anazi

Contributors
Reza Rahbarghazi, Jafar Rezaie, Morteza Heidarzadeh, Mehdi Hassanpour, Hassan Amini, Mahdi Ahmadi, Elhameh Shokrollahi, Pawan Kumar Gupta, Charan Thej, Gianandrea Pasquinelli, Carmen Ciavarella, Song Guo Zheng, Valerie Vanneaux, Khalid Ahmed Al-Anazi, Asma Al-Jasser, Waleed Al-Anazi

First published in London, United Kingdom, 2020 by IntechOpen
IntechOpen is the global imprint of INTECHOPEN LIMITED, registered in England and Wales, registration number: 11086078, 7th floor, 10 Lower Thames Street, London, EC3R 6AF, United Kingdom
Printed in Croatia

British Library Cataloguing-in-Publication Data
A catalogue record for this book is available from the British Library

Additional hard and PDF copies can be obtained from orders@intechopen.com

Update on Mesenchymal and Induced Pluripotent Stem Cells
Edited by Khalid Ahmed Al-Anazi
p. cm.
Print ISBN 978-1-78923-807-5
Online ISBN 978-1-78923-808-2
eBook (PDF) ISBN 978-1-83880-526-5

For EU product safety concerns:
IN TECH d.o.o., Prolaz Marije Krucifikse Kozulić 3, 51000 Rijeka, Croatia, info@intechopen.com or visit our website at intechopen.com.

Meet the editor

Dr Khalid Ahmed Al-Anazi is a consultant hemato-oncologist and the Chairman of the Department of Adult Hematology and Hematopoietic Stem Cell Transplantation (HSCT) at King Fahad Specialist Hospital (KFSH) in Dammam, Saudi Arabia.
He graduated from the College of Medicine, King Saud University (KSU) in Riyadh in 1986. After receiving his Boards in Internal Medicine, he did his fellowship training in Clinical Hematology and HSCT at King's College Hospital, University of London, U.K. He has 4 years of experience in internal medicine and 27 years of experience in adult clinical hematology and HSCT at Riyadh Armed Forces Hospital then King Faisal Specialist Hospital and Research Centre in Riyadh, King Khalid University Hospital (KKUH) and College of Medicine, KSU in Riyadh and KFSH in Dammam, Saudi Arabia.
He received the award for the best teacher in the Department of Medicine at the College of Medicine and KKUH in Riyadh in 2014. He established the adult HSCT program at KFSH in Dammam in 2010. He has more than 95 publications including retrospective studies, review articles, book chapters, and electronic books and he is a reviewer for 25 international medical journals. He is the Editor-in-Chief of the following journals: *Stem Cell Therapy and Transplantation*, *Stem Cell Biology and Transplantation*, and *Molecular Genetics and Medicine* in addition to being Associate Editor for 26 other medical journals in HSCT, hematology, cancer, and infectious diseases.

Contents

Preface

Since the first description or discovery of mesenchymal stem cells and induced pluripotent stem cells in the 1960s and 2006 respectively, these types of stem cells have witnessed enormous progress not only in their manufacture or preparation but also in their clinical utilization to treat several medical diseases. Many experts in the field believe that, in the near future, these specific types of stem cell may reshape the field of medical therapeutics and may eventually become potentially curative for several chronic and intractable medical conditions.

The book presents the latest information on both stem cell types. It is divided into three sections that cover several topics including: update on both cell types with a focus on their clinical utilizations; the use of various types of mesenchymal stem cells in the treatment of infectious diseases, osteoarthritis, and immunological disorders; in addition to the nature as well as the angiogenic paracrine properties of mesenchymal stem cells. Each chapter was written by scientists and clinicians with expertise in the field.

I would like to thank the authors for their valuable contributions, as well as the publishing manager Ms. Sara Debeuc and IntechOpen staff for their remarkable efforts that ultimately made this book project a reality.

Khalid A. Al-Anazi
Department of Hematology and Hematopoietic Stem Cell Transplantation,
Oncology Center, King Fahad Specialist Hospital,
Dammam, Saudi Arabia

Section 1

Mesenchymal and Induced Pluripotent Stem Cells - General Outline

Chapter 1

Introductory Chapter: Update on Mesenchymal and Induced Pluripotent Stem Cells

Khalid Ahmed Al-Anazi

1. Introduction

Stem cells are a subset of biological cells in the human body that are capable of self-renewal, tissue repair, differentiation, and division into different cell lineages [1–3]. Based on their origin and potency, stem cells are divided into either (1) embryonic and adult (non-embryonic) stem cells or (2) unipotent, oligopotent, totipotent, multipotent, and pluripotent stem cells [1, 2, 4, 5]. Multipotent or adult stem cells include mesenchymal stem cells (MSCs), while pluripotent stem cells include embryonic stem cells (ESCs) and induced pluripotent stem cells (iPSCs) [5].

2. MSCs

MSCs are heterogeneous, non-hematopoietic, adult multipotent stromal progenitor cells that are capable of self-renewal and differentiation into multiple lineages and various cell types [6–12]. They were first described in the 1960s by Alexander Friedenstein [7, 8, 10, 13]. They can be isolated from the bone marrow (BM), peripheral blood, umbilical cord blood, amniotic fluid, placenta, adipose tissue (AT), dental pulp, palatal tonsil, synovial fluid, salivary glands, as well as liver, lung, skin, and skeletal muscle tissues [6–13]. The main source of MSCs is the BM although MSCs constitute only a small fraction of the total number of cells populating the BM [7, 9–11].

MSCs have certain distinguishing features: being plastic adherent and ability of differentiation into osteoblasts, adipocytes, and chondrocytes, in addition to having characteristic surface markers [6–8, 10, 11, 13, 14]. On flow cytometry, they are characteristically positive for CD105, CD73, and CD90 and negative for CD45, CD34, CD11b, CD14, CD19, CD79a, and HLA-DR [6–8, 10, 11, 13]. However, several studies have shown that MSCs obtained from BM, AT, and other sources do express CD34 surface markers [9, 15–18]. MSCs can be seen in abundant numbers in the circulation under the following circumstances: stem cell mobilization with growth factors, tissue injuries, stroke, hypoxia, and inflammatory conditions [9, 19–24]. Despite the efforts made over the last five decades including identification of nine transcriptional factors, little is known about the molecular basis underlying the stemness of MSCs, and it is still unclear whether the recently discovered genes regulate stemness or only differentiation of MSCs [12].

MSCs have immunomodulatory and immunosuppressive properties that enable them to have several therapeutic and clinical applications, which include the enhancement of engraftment as well as prevention and treatment of graft versus

 IntechOpen

host disease (GVHD) in recipients of allogeneic hematopoietic stem cell transplantation (HSCT); treatment of several autoimmune disorders such as systemic lupus erythematosus, rheumatoid arthritis, systemic sclerosis, type I diabetes mellitus, and Crohn's disease; role in regenerative medicine and tissue repair including treatment of myocardial ischemia, myocardial infarction, cardiac dysfunction, dilated cardiomyopathy, chronic non-healing wounds, critical limb ischemia, liver injury, spinal cord injuries, as well as macular degeneration, corneal reconstruction, and transplantation; neurological disorders such as multiple sclerosis and amyotrophic lateral sclerosis; bone and cartilage diseases such as osteogenesis imperfecta; and treatment of various infections and acute respiratory distress syndrome [6, 7, 11, 25–27].

MSCs are major constituents of hematopoietic stem cell niche which is a highly complex and dynamic microenvironment of the BM [28]. Leptin receptor (LepR) is a marker that enriches BM-MSCs, and LepR+ cells in the BM are a major source of bone, cartilage, and adipocytes [29]. The exosome secretome of BM-MSCs regulates stem cell maintenance and their regenerative potential, and this BM-derived secretome will be critical to the future development of therapeutic strategies for oncologic diseases and regenerative medicine [30]. Apparently, MSCs are the masters of survival and clonality as they communicate with diverse immune cells and interact with other cellular components of the BM microenvironment as well as with normal cells, leukemic stem cells, and progenitor cells [31]. The main functions of MSCs include formation of hematopoietic microenvironment, modulation of the activity of the immune system, and regulating cell trafficking [32]. When stimulated by specific signals, MSCs can be released from BM niche into circulation and can be recruited to the target tissues where they undergo in situ differentiation and contribute to tissue regeneration and homeostasis [33]. The efficacy of MSCs is linked to their immunosuppressive and anti-inflammatory properties primarily due to the release of soluble factors [34].

The putative roles of BM-MSCs during infection are detection of pathogens; activation of host immune response; elimination of pathogens; induction of proinflammatory gradients; and modulation of proinflammatory host immune response [6, 7]. Examples of the immunoregulatory properties of MSCs include inhibition of differentiation of monocytes to dendritic cells (DCs), alteration of cytokine profile of DCs, induction of tolerant phenotypes of naïve and effector T cells, inhibition of antibody production by B cells, and suppression of natural killer (NK) cell proliferation and NK-mediated cytotoxicity [35]. BM-MSCs may augment antimicrobial responses, abridge proinflammatory and damage responses, and ameliorate injury caused by the host defense to the pathogen [6, 7]. BM-MSCs appear to function as a critical fulcrum providing balance by promoting pathogen clearance during the initial inflammatory response and suppressing inflammation to preserve host integrity and facilitate tissue repair [6].

MSCs could potentially be involved at multiple levels in host defense by mobilizing immune effector cells and modulation of proinflammatory immune responses so as to minimize the tissue damage induced by inflammation [6, 36]. The immunomodulatory properties of MSCs are mediated by both: cell to cell interaction and the secreted cytokines [36, 37]. BM-MSCs may protect against infectious challenge either by direct effects on the pathogens or through indirect effects on the host [6]. On the other hand, certain types of MSCs, particularly placenta-derived MSCs and fetal membrane-derived MSCs, are highly susceptible to herpes viruses including varicella zoster virus [7, 38].

Studies have shown that several types of stem cells including BM-MSCs and neural stem cells can cross the blood brain barrier and reach tumors localized in the brain such as glioblastoma multiforme as well as ischemic areas and injured sites in the brain and engraft there. Hence, MSCs can be used as means of cellular carriers or Trojan horses to deliver cytotoxic genes or therapeutic agents for brain tumors, and they can be used to exert their therapeutic and regenerative effects in the brain [39–43]. In cancer, MSCs are a double-edged sword as they can exert stimulatory effects on tumor development, while they can have inhibitory effects on cancer cell growth and metastases [44]. MSCs have anticancer properties, and they can be engineered or modified to become carriers of suicide genes, employed as carriers of anti-angiogenesis factors, and utilized to target cancer stem cells [45–47]. MSCs have recently been engineered to express antiproliferative, anti-apoptotic, and antiangiogenic agents that specifically target different types of solid tumors [45].

The capacity of MSCs to proliferate and differentiate into other cells in addition to their ability to release biomolecules such as cytokines, growth factors, and microvesicles that have anti-inflammatory, immunomodulatory, anti-fibrogenic, and trophic functions make MSCs ideal candidates to function as delivery platform for cellular and gene therapies [48, 49]. Consequently, clinical trials incorporating the utilization of MSCs in the treatment of immune-related diseases have rapidly evolved after reports from preclinical studies confirming their safety and efficacy [49]. Recently, scientists have established several strategies to generate highly functional AT-derived MSCs and these include preconditioning of AT-MSCs with various stimulants and inflammatory agents; genetic manipulation of AT-MSCs; modification of culture conditions with three-dimensional aggregate formation and hypoxic culture; and proper utilization of exosome and extracellular vesicles (ECVs) that are secreted by AT-MSCs [50, 51]. Also, the main focus has recently shifted from studying differentiation of MSCs to studying their paracrine properties such as the release of ECVs that contain numerous micro-RNAs (miRNAs) including regulatory miRNAs and the production of multiple bioactive proteins and compounds that regulate MSC differentiation [52]. Hence, soluble elements derived from MSCs including ECVs have recently been proposed as a cell-free alternative for various therapies on the clinical side [51].

The combination of MSCs and tissue engineering technology can enhance the immunoregulatory properties of MSCs, and this will ultimately lead to further expansion of their utilization in regenerative medicine [53]. Tissue engineering strategies such as the use of various types of stem cells, scaffolds, medical devices, gene therapy, and nanotopography have resulted in progressing the translation of basic research towards clinical therapeutics [54, 55]. Despite the remarkable progress in MSC therapies, sufficient data on the biodistribution of MSCs, cellular and molecular structures of their target cells, and mechanisms by which MSCs reach these targets are still lacking [56]. Also, several obstacles need to be overcome before the utilization of specific types of MSCs in tissue engineering becomes a routine practice in the clinical arena [57]. Currently, human MSCs are generated through conventional static adherent cultures in the presence of fetal bovine serum or human-sourced supplements. Unfortunately, these methods are not ideal procedures to meet the future expectations of quality-assured human MSCs for clinical therapies in humans [58]. Additionally, having substantial gaps in our knowledge of the biology and therapeutic efficacy of MSCs presents major challenge to their sustainable implementation in clinical medicine [59]. Thus, optimizing the bioprocess to generate human MSCs and their products will improve efficacy and safety of stem cell therapies [58]. Also, improving the cultural environment of MSCs and selecting the appropriate scaffolds and induction factors are essential in improving the outcome of MSC-based tissue engineering [60].

3. iPSCs

Human iPSCs resemble human ESCs in many aspects including morphology, proliferation, differentiation potential, and pluripotency markers, but the epigenetic characteristics of human iPSCs are rather distinct [1, 2, 5, 61]. Although the utilization of iPSCs can avoid the obstacles and ethical concerns that limit the use of human ESCs, clinical application of human iPSCs still has a number of disadvantages that include chromosomal instability and tumorigenic potential, thus raising questions about the safety of their clinical utilization, and low reprogramming efficiency in addition to other concerns about their reproducibility for laboratory applications in disease modelling and drug screening [1, 3, 5, 61, 62].

In 2006, Takahashi and Yamanaka were the first scientists to generate mouse iPSCs from dermal fibroblasts through retroviral-mediated ectopic expression of the four genes: OCT4, SOX2, KLF4, and c-MYC [1, 3, 4, 63]. Since this discovery, iPSCs have been used in many research and clinical trials, including disease modelling; drug toxicity as well as drug discovery; and regenerative medicine [3–5]. Reprogramming of iPSCs should have the following crucial requirements: species such as human or mouse; cell type such as blood cell or fibroblast; factor, drug, chemical, or other protein molecules such as miRNA, DNA modifying agent, NANOG, or LIN28; vector such as retrovirus or lentivirus; and disease with specific genetic mutation [1, 4, 5, 64].

Human iPSCs have revolutionized the field of human disease modelling with an enormous potential to serve as paradigm shifting platforms for preclinical trials, personalized clinical diagnosis, and personalized drug therapy [65]. During the last 13 years, significant developments and remarkable progress have been achieved in enhancing reprogramming techniques and their efficacy, increasing safety of derived iPSCs, and developing different delivery methods [61, 62]. The ability to generate iPSCs from human somatic cells provides tremendous promises and opportunities in basic research and regenerative medicine and can provide a wide range of applications including cell-based therapies, drug screening, and disease modelling [61, 66].

The capacity of human iPSCs to retain patient-specific genomic, transcriptomic, proteomic, metabolomic, and other visualized big data information makes it possible to extend their applications beyond disease modelling into the field of personalized medicine which encompasses the adoption of novel prevention and treatment strategies based on individual variability [65]. The emergence of modern iPSC technology, with the capacity of these stem cells to undergo unlimited self-renewal and differentiation into any type of cell, has a great potential to advance translational applications including stem cell therapies and the generation of large-scale collections of cell lines for research purposes [67]. Recently, genomic editing technologies have been applied to correct the mutations in disease-specific iPSCs to create gene-corrected iPSCs that can be utilized in autologous stem cell-based therapies [64]. Nowadays, patient-specific iPSCs can be obtained by reprogramming of adult somatic cells by ectopic expression of pluripotency-associated transcription factors including OCT4, SOX2, KLF4, and c-MYC [64]. The availability of precisely generated iPSC-derived functional cells to replace or repair damaged tissues or organs will likely affect therapies of hematopoietic disorders and facilitate treatment of neurological, cardiovascular, hepatic, and retinal diseases and possibly diabetes mellitus [67]. Additionally, patient-specific iPSCs can bypass certain limitations of ESCs such as ethical concerns and immunological rejection [64]. The first clinical trial on cell-based therapy using iPSCs derived from patients to treat blindness started in Japan in September 2014 [67].

4. iPSC-MSCs and conclusion

MSCs derived from iPSCs (iPSC-MSCs) exhibit higher proliferation rate and less senescence than BM-MSCs, and thus the former cells are emerging as an attractive therapeutic option for obtaining a substantial population of stem cells in a sustained manner for applications in regenerative medicine [68, 69]. Several studies using human iPSC-MSCs and their exosomes in human and animal studies have shown that transplantation of these cells can produce protection of the liver against hepatic ischemia; reduction in the volume of brain infarction and preservation of neurological function after acute intracranial hemorrhage; prevention of osteonecrosis of femoral head by promotion of local angiogenesis and prevention of bone loss; facilitation of cutaneous wound healing by promotion of collagen synthesis and angiogenesis; and modulation of differentiation and function of DCs in order to support their clinical application in DC-mediated immune disorders [69–73]. Thus, MSCs and iPSCs may reshape the future of medical therapeutics and may eventually become curative for several chronic and intractable medical illnesses [2, 4, 5].

Author details

Khalid Ahmed Al-Anazi
Department of Hematology and Hematopoietic Stem Cell Transplantation, Oncology Center, King Fahad Specialist Hospital, Dammam, Saudi Arabia

*Address all correspondence to: kaa_alanazi@yahoo.com

References

[1] Menon S, Shailendra S, Renda A, Longaker M, Quarto N. An overview of direct somatic reprogramming: The ins and outs of iPSCs. International Journal of Molecular Sciences. 2016;**17**(1). pii: E141. DOI: 10.3390/ijms17010141

[2] Al-Anazi KA. Stem cell treatments may reshape the future of medical therapeutics. Journal of Stem Cell Biology and Transplantation. 2016;**1**(1:1). DOI: 10.21767/2575-7725.100001

[3] Diecke S, Jung SM, Lee J, Ju JH. Recent technological updates and clinical applications of induced pluripotent stem cells. The Korean Journal of Internal Medicine. 2014;**29**(5):547-557. DOI: 10.3904/kjim.2014.29.5.547

[4] Singh VK, Kalsan M, Kumar N, Saini A, Chandra R. Induced pluripotent stem cells: Applications in regenerative medicine, disease modeling, and drug discovery. Frontiers in Cell and Development Biology. 2015;**3**:2. DOI: 10.3389/fcell.2015.00002

[5] Al-Anazi KA. Induced pluripotent stem cells and their future therapeutic applications in hematology. Stem Cell Research and Therapy. 2015;**5**:258. DOI: 10.4172/2157-7633.1000258

[6] Auletta JJ, Deans RJ, Bartholomew AM. Emerging roles for multipotent, bone marrow-derived stromal cells in host defense. Blood. 2012;**119**(8):1801-1809. DOI: 10.1182/blood-2011-10-384354

[7] Al-Anazi KA, Al-Jasser AM. Mesenchymal stem cells-their antimicrobial effects and their promising future role as novel therapies of infectious complications in high risk patients. In: Demirer T, editor. Progress in Stem Cell Transplantation. IntechOpen; 2015. DOI: 10.5772/60640

[8] Abdal Dayem A, Lee SB, Kim K, Lim KM, Jeon TI, Seok J, et al. Production of mesenchymal stem cells through stem cell reprogramming. International Journal of Molecular Sciences. 2019;**20**(8). pii: E1922. DOI: 10.3390/ijms20081922

[9] Al-Anazi KA, Bakhit K, Al-Sagheir A, AlHashmi H, Abdulbaqi M, Al-Shibani Z, et al. Cure of insulin-dependent diabetes mellitus by an autologous hematopoietic stem cell transplantation performed to control multiple myeloma in a patient with chronic renal failure on regular hemodialysis. Journal of Stem Cell Biology and Transplantation. 2017;**1**(2:11). DOI: 10.21767/2575-7725.100011

[10] Bobis S, Jarocha D, Majka M. Mesenchymal stem cells: Characteristics and clinical applications. Folia Histochemica et Cytobiologica. 2006;**44**(4):215-230

[11] Kim N, Cho SG. Clinical applications of mesenchymal stem cells. Korean Journal of Internal Medicine. 2013;**28**(4):387-402. DOI: 10.3904/kjim.2013.28.4.387

[12] Liu TM. Stemness of mesenchymal stem cells. Preliminary study. Journal of Stem Cell Therapy and Transplantation. 2017;**1**:071-073. DOI: 10.29328/journal.jsctt.1001008

[13] Squillaro T, Peluso G, Galderisi U. Clinical trials with mesenchymal stem cells: An update. Cell Transplantation. 2016;**25**(5):829-848. DOI: 10.3727/096368915X689622

[14] Dominici M, Le Blanc K, Mueller I, Slaper-Cortenbach I, Marini F, Krause D, et al. Minimal criteria for defining multipotent mesenchymal stromal cells. The International Society for Cellular

Therapy position statement. Cytotherapy. 2006;**8**(4):315-317. DOI: 10.1080/14653240600855905

[15] Lin CS, Ning H, Lin G, Lue TF. Is CD34 truly a negative marker for mesenchymal stromal cells? Cytotherapy. 2012;**14**(10):1159-1163. DOI: 10. 3109/14653249.2012.729817

[16] Sidney LE, Branch MJ, Dunphy SE, Dua HS, Hopkinson A. Concise review: Evidence for CD34 as a common marker for diverse progenitors. Stem Cells. 2014;**32**(6):1380-1389. DOI: 10.1002/stem.1661

[17] Stzepourginski I, Nigro G, Jacob JM, Dulauroy S, Sansonetti PJ, Eberl G, et al. CD34+ mesenchymal cells are a major component of the intestinal stem cells niche at homeostasis and after injury. Proceedings of the National Academy of Sciences of the United States of America. 2017;**114**(4):E506-E513. DOI: 10.1073/pnas.1620059114

[18] Eto H, Ishimine H, Kinoshita K, Watanabe-Susaki K, Kato H, Doi K, et al. Characterization of human adipose tissue-resident hematopoietic cell populations reveals a novel macrophage subpopulation with CD34 expression and mesenchymal multipotency. Stem Cells and Development. 2013;**22**(6):985-997. DOI: 10.1089/scd.2012.0442

[19] Alvarez P, Carrillo E, Vélez C, Hita-Contreras F, Martínez-Amat A, Rodríguez-Serrano F, et al. Regulatory systems in bone marrow for hematopoietic stem/progenitor cells mobilization and homing. BioMed Research International. 2013;**2013**:312656. DOI: 10.1155/2013/312656

[20] Rochefort GY, Delorme B, Lopez A, Hérault O, Bonnet P, Charbord P, et al. Multipotential mesenchymal stem cells are mobilized into peripheral blood by hypoxia. Stem Cells. 2006;**24**(10):2202-2208

[21] Lund TC, Tolar J, Orchard PJ. Granulocyte colony-stimulating factor mobilized CFU-F can be found in the peripheral blood but have limited expansion potential. Haematologica. 2008;**93**(6):908-912. DOI: 10.3324/haematol.12384

[22] Gilevich IV, Fedorenko TV, Pashkova IA, Porkhanov VA, Chekhonin VP. Effects of growth factors on mobilization of mesenchymal stem cells. Bulletin of Experimental Biology and Medicine. 2017;**162**(5):684-686. DOI: 10.1007/s10517-017-3687-0

[23] Xu L, Li G. Circulating mesenchymal stem cells and their clinical implications. Journal of Orthopaedic Translation. 2014;**2**(1):1-7. DOI: 10.1016/j.jot.2013.11.002

[24] Koning JJ, Kooij G, de Vries HE, Nolte MA, Mebius RE. Mesenchymal stem cells are mobilized from the bone marrow during inflammation. Frontiers in Immunology. 2013;**4**:49. DOI: 10.3389/fimmu.2013.00049

[25] Ding SSL, Subbiah SK, Khan MSA, Farhana A, Mok PL. Empowering mesenchymal stem cells for ocular degenerative disorders. International Journal of Molecular Sciences. 2019;**20**(7). pii: E1784. DOI: 10.3390/ijms20071784

[26] Mansoor H, Ong HS, Riau AK, Stanzel TP, Mehta JS, Yam GH. Current trends and future perspective of mesenchymal stem cells and exosomes in corneal diseases. International Journal of Molecular Sciences. 2019;**20**(12). pii: E2853. DOI: 10.3390/ijms20122853

[27] Leyendecker A Jr, Pinheiro CCG, Amano MT, Bueno DF. The use of human mesenchymal stem cells as therapeutic agents for the in vivo treatment of immune-related diseases: A systematic review. Frontiers in Immunology. 2018;**9**:2056. DOI: 10.3389/fimmu.2018.02056

[28] Pinho S, Lacombe J, Hanoun M, Mizoguchi T, Bruns I, Kunisaki Y, et al. PDGFRα and CD51 mark human nestin+ sphere-forming mesenchymal stem cells capable of hematopoietic progenitor cell expansion. Journal of Experimental Medicine. 2013;**210**(7):1351-1367. DOI: 10.1084/jem.20122252

[29] Zhou BO, Yue R, Murphy MM, Peyer JG, Morrison SJ. Leptin-receptor-expressing mesenchymal stromal cells represent the main source of bone formed by adult bone marrow. Cell Stem Cell. 2014;**15**(2):154-168. DOI: 10.1016/j.stem.2014.06.008

[30] Eltoukhy HS, Sinha G, Moore CA, Gergues M, Rameshwar P. Secretome within the bone marrow microenvironment: A basis for mesenchymal stem cell treatment and role in cancer dormancy. Biochimie. 2018;**155**:92-103. DOI: 10.1016/j.biochi.2018.05.018

[31] Pleyer L, Valent P, Greil R. Mesenchymal stem and progenitor cells in normal and dysplastic hematopoiesis-masters of survival and clonality? International Journal of Molecular Sciences. 2017;**17**(7). pii: E1009. DOI: 10.3390/ijms17071009

[32] Shi C. Recent progress toward understanding the physiological function of bone marrow mesenchymal stem cells. Immunology. 2012;**136**(2):133-138. DOI: 10.1111/j.1365-2567.2012.03567.x

[33] Liu ZJ, Zhuge Y, Velazquez OC. Trafficking and differentiation of mesenchymal stem cells. Journal of Cellular Biochemistry. 2009;**106**(6): 984-991. DOI: 10.1002/jcb.22091

[34] De Luca L, Trino S, Laurenzana I, Lamorte D, Caivano A, Del Vecchio L, et al. Mesenchymal stem cell derived extracellular vesicles: A role in hematopoietic transplantation? International Journal of Molecular Sciences. 2017;**18**(5). pii: E1022. DOI: 10.3390/ijms18051022

[35] Thanunchai M, Hongeng S, Thitithanyanont A. Mesenchymal stromal cells and viral infection. Stem Cells International. 2015;**2015**:860950. DOI: 10.1155/2015/860950

[36] Castro-Manrreza ME, Montesinos JJ. Immunoregulation by mesenchymal stem cells: Biological aspects and clinical applications. Journal of Immunology Research. 2015;**2015**:394917. DOI: 10.1155/2015/394917

[37] Kyurkchiev D, Bochev I, Ivanova-Todorova E, Mourdjeva M, Oreshkova T, Belemezova K, et al. Secretion of immunoregulatory cytokines by mesenchymal stem cells. World Journal of Stem Cells. 2014;**6**(5):552-570. DOI: 10.4252/wjsc.v6.i5.552

[38] Avanzi S, Leoni V, Rotola A, Alviano F, Solimando L, Lanzoni G, et al. Susceptibility of human placenta derived mesenchymal stromal/stem cells to human herpesviruses infection. PLoS ONE. 2013;**8**(8):e71412. DOI: 10.1371/journal.pone.0071412

[39] Abdi Z, Eskandary H, Nematollahi-Mahani SN. Effects of two types of human cells on outgrowth of human glioma in rats. Turkish Neurosurgery. 2018;**28**(1):19-28. DOI: 10.5137/1019-5149.JTN.18697-16.1

[40] Dong HJ, Li G, Meng HP, Shang CZ, Luo Y, Wen G, et al. How can mesenchymal stem cells penetrate the blood brain barrier? Turkish Neurosurgery. 2018;**28**(6):1013-1014. DOI: 10.5137/1019-5149.JTN.22639-18.1

[41] Conaty P, Sherman LS, Naaldijk Y, Ulrich H, Stolzing A, Rameshwar P. Methods of mesenchymal

stem cell homing to the blood-brain barrier. Methods in Molecular Biology. 2018;**1842**:81-91. DOI: 10.1007/978-1-4939-8697-2_6

[42] Liu L, Eckert MA, Riazifar H, Kang DK, Agalliu D, Zhao W. From blood to the brain: Can systemically transplanted mesenchymal stem cells cross the blood-brain barrier? Stem Cells International. 2013;**2013**:435093. DOI: 10.1155/2013/435093

[43] Christodoulou I, Goulielmaki M, Devetzi M, Panagiotidis M, Koliakos G, Zoumpourlis V. Mesenchymal stem cells in preclinical cancer cytotherapy: A systematic review. Stem Cell Research & Therapy. 2018;**9**(1):336. DOI: 10.1186/s13287-018-1078-8

[44] Lee HY, Hong IS. Double-edged sword of mesenchymal stem cells: Cancer-promoting versus therapeutic potential. Cancer Science. 2017;**108**(10):1939-1946. DOI: 10.1111/cas.13334

[45] Nowakowski A, Drela K, Rozycka J, Janowski M, Lukomska B. Engineered mesenchymal stem cells as an anti-cancer trojan horse. Stem Cells and Development. 2016;**25**(20):1513-1531. DOI: 10.1089/scd.2016.0120

[46] Zhang L, Su XS, Ye JS, Wang YY, Guan Z, Yin YF. Bone marrow mesenchymal stem cells suppress metastatic tumor development in mouse by modulating immune system. Stem Cell Research & Therapy. 2015;**6**:45. DOI: 10.1186/s13287-015-0039-8

[47] Fakiruddin KS, Ghazalli N, Lim MN, Zakaria Z, Abdullah S. Mesenchymal stem cell expressing TRAIL as targeted therapy against sensitised tumor. International Journal of Molecular Sciences. 2018;**19**(8). pii: E2188. DOI: 10. 3390/ijms19082188

[48] D'souza N, Rossignoli F, Golinelli G, Grisendi G, Spano C, Candini O, et al. Mesenchymal stem/stromal cells as a delivery platform in cell and gene therapies. BMC Medicine. 2015;**13**:186. DOI: 10.1186/s12916-015-0426-0

[49] Wang M, Yuan Q, Xie L. Mesenchymal stem cell-based immunomodulation: Properties and clinical application. Stem Cells International. 2018;**2018**:3057624. DOI: 10.1155/2018/3057624

[50] Seo Y, Shin TH, Kim HS. Current strategies to enhance adipose stem cell function: An update. International Journal of Molecular Sciences. 2019;**20**(15). pii: E3827. DOI: 10.3390/ijms20153827

[51] Najar M, Bouhtit F, Melki R, Afif H, Hamal A, Fahmi H, et al. Mesenchymal stromal cell-based therapy: New perspectives and challenges. Journal of Clinical Medicine. 2019;**8**(5). pii: E626. DOI: 10.3390/jcm8050626

[52] Lukomska B, Stanaszek L, Zuba-Surma E, Legosz P, Sarzynska S, Drela K. Challenges and controversies in human mesenchymal stem cell therapy. Stem Cells International. 2019;**2019**:9628536. DOI: 10.1155/2019/9628536

[53] Li H, Shen S, Fu H, Wang Z, Li X, Sui X, et al. Immunomodulatory functions of mesenchymal stem cells in tissue engineering. Stem Cells International. 2019;**2019**:9671206. DOI: 10.1155/2019/9671206

[54] Salmasi S, Kalaskar DM, Yoon WW, Blunn GW, Seifalian AM. Role of nanotopography in the development of tissue engineered 3D organs and tissues using mesenchymal stem cells. World Journal of Stem Cells. 2015;7(2):266-280. DOI: 10.4252/wjsc.v7.i2.266

[55] Fang L, Shi XF, Sun HY, Li YX, Xiao FJ, Wang H, et al. Effects of exosomes derived from miR-486 gene modified umbilical cord

mesenchymal stem cells on biological characteristics of rat cardiomyocytes. Zhongguo Shi Yan Xue Ye Xue Za Zhi. 2018;**26**(5):1531-1537. DOI: 10.7534/j.issn.1009-2137.2018.05.045

[56] Leibacher J, Henschler R. Biodistribution, migration and homing of systemically applied mesenchymal stem/stromal cells. Stem Cell Research & Therapy. 2016;7:7. DOI: 10.1186/s13287-015-0271-2

[57] Nancarrow-Lei R, Mafi P, Mafi R, Khan W. A systemic review of adult mesenchymal stem cell sources and their multilineage differentiation potential relevant to musculoskeletal tissue repair and regeneration. Current Stem Cell Research & Therapy. 2017;**12**(8):601-610. DOI: 10.2174/1574888X12666170608124303

[58] Panchalingam KM, Jung S, Rosenberg L, Behie LA. Bioprocessing strategies for the large-scale production of human mesenchymal stem cells: A review. Stem Cell Research & Therapy. 2015;**6**:225. DOI: 10.1186/s13287-015-0228-5

[59] Schäfer R. Advanced cell therapeutics are changing the clinical landscape: Will mesenchymal stromal cells be a part of it? BMC Medicine. 2019;**17**(1):53. DOI: 10.1186/s12916-019-1289-6

[60] Han Y, Li X, Zhang Y, Han Y, Chang F, Ding J. Mesenchymal stem cells for regenerative medicine. Cell. 2019;**8**(8). pii: E866. DOI: 10.3390/cells8080886

[61] Brouwer M, Zhou H, Nadif Kasri N. Choices for induction of pluripotency: Recent developments in human induced pluripotent stem cell reprogramming strategies. Stem Cell Reviews and Reports. 2016;**12**(1):54-72. DOI: 10.1007/s12015-015-9622-8

[62] Hong SG, Dunbar CE, Winkler T. Assessing the risks of genotoxicity in the therapeutic development of induced pluripotent stem cells. Molecular Therapy. 2013;**21**(2):272-281. DOI: 10.1038/mt.2012.255

[63] Wu Y-Y, Chiu F-L, Yeh C-S, Kuo H-C. Opportunities and challenges for the use of induced pluripotent stem cells in modelling neurodegenerative disease. Open Biology. 2019;**9**:180177. DOI: 10.1098/rsob.180177

[64] Wattanapanitch M. Recent updates on induced pluripotent stem cells in hematological disorders. Stem Cells International. 2019;**2019**:5171032. DOI: 10.1155/2019/5171032

[65] Matsa E, Ahrens JH, Wu JC. Human induced pluripotent stem cells as a platform for personalized and precision cardiovascular medicine. Physiological Reviews. 2016;**96**(3):1093-1126. DOI: 10.1152/physrev.00036.2015

[66] Singh VK, Kumar N, Kalsan M, Saini A, Chandra R. Mechanism of induction: Induced pluripotent stem cells (iPSCs). Journal of Stem Cells. 2015;**10**(1):43-62

[67] Zhou H, Martinez H, Sun B, Li A, Zimmer M, Katsanis N, et al. Rapid and efficient generation of transgene-free iPSC from a small volume of cryopreserved blood. Stem Cell Reviews and Reports. 2015;**11**(4):652-665. DOI: 10.1007/s12015-015-9586-8

[68] Sabapathy V, Kumar S. hiPSC-derived iMSCs: NextGen MSCs as an advanced therapeutically active cell resource for regenerative medicine. Journal of Cellular and Molecular Medicine. 2016;**20**(8):1571-1588. DOI: 10.1111/jcmm.12839

[69] Gao WX, Sun YQ, Shi J, Li CL, Fang SB, Wang D, et al. Effects of mesenchymal stem cells from human induced pluripotent stem cells on differentiation, maturation, and function of dendritic cells. Stem Cell

Research & Therapy. 2017;**8**(1):48. DOI: 10.1186/s13287-017-0499-0

[70] Du Y, Li D, Han C, Wu H, Xu L, Zhang M, et al. Exosomes from human-induced pluripotent stem cell-derived mesenchymal stromal cells (hiPSC-MSCs) protect liver against hepatic ischemia/reperfusion injury via activating sphingosine kinase and sphingosine-1-phosphate signaling pathway. Cellular Physiology and Biochemistry. 2017;**43**(2):611-625. DOI: 10.1159/000480533

[71] Chen KH, Lin KC, Wallace CG, Li YC, Shao PL, Chiang JY, et al. Human induced pluripotent stem cell-derived mesenchymal stem cell therapy effectively reduced brain infarct volume and preserved neurological function in rat after acute intracranial hemorrhage. American Journal of Translational Research. 2019;**11**(9):6232-6248

[72] Liu X, Li Q, Niu X, Hu B, Chen S, Song W, et al. Exosomes secreted from human-induced pluripotent stem cell-derived mesenchymal stem cells prevent osteonecrosis of the femoral head by promoting angiogenesis. International Journal of Biological Sciences. 2017;**13**(2):232-244. DOI: 10.7150/ijbs.16951

[73] Zhang J, Guan J, Niu X, Hu G, Guo S, Li Q, et al. Exosomes released from human induced pluripotent stem cells-derived MSCs facilitate cutaneous wound healing by promoting collagen synthesis and angiogenesis. Journal of Translational Medicine. 2015;**13**:49. DOI: 10.1186/s12967-015-0417-0

Chapter 2

Induced Pluripotent Stem Cells for Clinical Use

Valérie Vanneaux

Abstract

The use of induced pluripotent stem cells (iPSCs) represents a great promise in regenerative medicine. So far, several clinical trials are underway and preliminary results are promising with the human embryonic stem cells, their non-reprogrammed counterparts. The experience of the clinical use of iPSC derivatives is extremely limited because of several major safety concerns, but many technological advances in the field of iPSC generate high expectations in the near future to develop new clinical trials with an adapted level of patient safety. New guidelines and several recommendations are edited by researchers and regulatory agencies to guarantee the safety of the iPSC products in a clinical context for medical applications. In this chapter, we review the clinical trials with induced pluripotent stem cells and the main factors affecting the safe translation of iPSC to the clinic and how to overcome these issues by standardization and to control the quality of the clinical-grade iPSC products.

Keywords: clinical-grade induced pluripotent stem cells, regenerative medicine, clinical trials, quality control

1. Introduction

Since the discovery of induced pluripotent stem cells (iPSCs) by Yamanaka and Takahashi in 2006, many expectations have emerged, and iPSCs have opened up a world of possibilities for new cell-based therapies in regenerative medicine [1]. In the domain of pluripotent stem cells, iPSCs are considered as equivalent to embryonic stem cells (ESCs), because of two intrinsic key properties: their indefinite proliferative capacities while preserving pluripotency and their capacity to differentiate into all known cell types. However, in contrast to ESCs, iPSCs can be generated without any controversial ethical issues, thus favoring their use in clinical settings. Last but not least, in an autologous approach of cell-based therapy, by using the patient's own cells as source for iPSC generation, one circumvents all the issues related to the immunological compatibility between the donor and receiver. This largely explains the tremendous enthusiasm engendered by iPSC discovery in the sphere of regenerative medicine during the last decade. In this review article, we provide an overview of the launched clinical trials with iPSC and the ongoing efforts to understand the risk related to safety of iPSC-derived cells, highlighting some of the problems that have to be overcome.

2. Clinical trials with iPSC

After over a decade of research on iPSC, and due to fast-track facilitating procedure in Japan, several clinical studies were launched. While the first clinical trial based on the human ESC started in 2010, taking advantage of the acquired extensive knowledge of ESC biology, despite their relatively recent discovery, the first clinical study based on the iPSC-derived retinal pigmented epithelium was authorized and conducted at the RIKEN Institute in Japan in 2014 [2]. A sheet of autologous iPSC-derived retinal cells were transplanted in a patient with eye-related macular degeneration (AMD). In 2015, the RIKEN Institute decided to suspend the study due to safety concerns on the cells of the second recruited patient [3]. Nonetheless, regarding the first transplanted patient, a 25-month follow-up revealed neither serious events, nor clinical signs of rejection. Moreover, the macular degeneration progress was delayed in the treated eye compared to the untreated eye. This result corroborated all the results obtained previously in the course of the ESC-based clinical studies, where no adverse events related to transplanted cells were observed. Still this problem induced a shift in the approach from patient-specific autologous to highly securized allogeneic iPSC lines. This study was resumed in 2017 and until now five patients with AMD have been treated with allogeneic iPSC-derived cells.

Since then, several clinical studies based on allogeneic iPSCs have been developed and approved. Until mid-2019, there have been nine ongoing clinical studies based on iPSC, mostly nationally approved in Japan, with four of them being approved in the first months of 2019, with indications including Parkinson's disease, AMD, severe cardiac failure, aplastic anemia, spinal cord injury and corneal stem cell deficiency. Furthermore, two private companies—Cynata Therapeutics, an Australian stem cell and regenerative medicine company, and Fate Therapeutics, an American clinical-stage biopharmaceutical company—have developed a line of products based on allogeneic human iPSC-derived cells. In Australia and United Kingdom, Cynata Therapeutics just concluded a phase I study using CYP-001, an iPSC-derived mesenchymoangioblast precursor administered intravenously in 15 patients with graft-versus-host disease (GVHD) occurring after an allogeneic hematopoietic stem cell transplant [4]. Currently, all patients treated so far have demonstrated at least a partial response, while no treatment-related serious adverse events or safety concerns have been observed. The product development activities of CYP-001 will be done in a phase II study in 2019 by Fujifilm in collaboration with Cynata Therapeutics. On its part, Fate Therapeutics received a first approval from Food and Drug Administration (FDA) in November 2018 to transplant an off-the-shelf iPSC-derived Natural Killer cell, FT-500, as cancer immunotherapy to treat solid tumors and for a second cell product derived from a genetically engineered iPSC, FT-516, in February 2019, for the treatment of relapsed/refractory hematologic malignancies. For the first product FT-500, all the three patients with advanced solid tumors have been treated with multiple doses of FT-500, 100 million cells per dose, and it has been well tolerated with no dose-limiting toxicities or adverse events [5].

Even though the first clinical studies have already been started, technical advances in iPSC biology have revealed that several factors could affect their safety for a larger range of medical applications, and should be taken into account for short- and long-term follow-up of patients. Two of the major concerns related to iPSC-based products are their potential tumorigenicity and immunogenicity. The scientific community is still continuing to elucidate the biological mechanisms underlying iPSC's immunogenicity and tumorigenicity and how to manage or overcome them.

3. iPSC safety

3.1 Tumorigenicity

The potential risk of tumorigenicity to patients from both teratomas and malignant tumors could arise if transplanted cells are contaminated with undifferentiated iPSC, or if transplanted cells have been genetically modified and become unstable during the *in vitro* production steps.

The major concern related to iPSC-based tumorigenicity is the reprogramming method. In the original cocktail of transcription factors developed by Yamanaka, somatic cells are transduced by retroviral vectors that become integrated into the genome of the host cells. Two of these factors—*c-Myc* and *klf4*—are potent oncogenes [6]. Subsequently, reports of tumorigenicity after transplantation of iPSC or iPSC-derived cells are not surprising. Thereby, teratoma formation could be induced by the undesired activation/suppression of essential host genes proximal to integration sites or by residual expression of reprogramming factors in the derived cells in animal model [7, 8]. With hindsight, there is evidence for the necessity to select a non-integrative method for reprogramming, a higher rate of genomic alterations occurring when human iPSCs are generated with viral vectors, compared to mRNA [7, 9]. Numerous studies, focused on the choice of reprogramming factors and methods of delivery, have developed various novel strategies to enhance the efficiency of reprogramming and reduce the potential risk of tumorigenicity. To circumvent this risk, human iPSCs have been generated by several "integration-free" methods, based on the use of viral vectors (adenoviral vectors and Sendai virus-based vectors) or non-viral vectors (piggyBac system, minicircle vector, and episomal vectors). Originally, the four transcription factors needed for complete cell reprograming were *c-myc*, *klf4*, *oct4* and *sox2* [1]. The pro-tumorigenic transcription factor *c-myc* has been found to be unnecessary for the reprogramming process, but the overall efficiency is decreased without it. Several strategies have been developed with the use of different transcription factors and/or replacement of *c-myc*, or the use of direct protein delivery and synthesized mRNA [10–12].

Furthermore, the tumorigenicity risk is often linked to the genetic instability of iPSC. Random genomic alterations are frequently observed in human iPSCs showing their intrinsic instability, essentially due to the massive genome remodeling, and probably also resulting from various mechanisms such as replicative stress, reactivation of the telomerase and metabolism modification from the oxidative to the glycolytic state. Epigenetic modifications may also contribute to iPSC variation due to residual epigenetic memories of the starting cell type [13]. The incomplete resetting of the non-CpG methylation patterns during reprogramming leads to a biased differential potential in certain cell types depending on the donor cell source [14, 15]. However, it has been shown that their residual epigenetic memory diminishes with the *in vitro* expansion over a period of time [16, 17]. As just mentioned, the selection of the donor cell type is of importance. Many human somatic cell types have been successfully reprogrammed. However, even if the use of different transcription factors, delivery methods and culture conditions does not facilitate any comparison, it is well known that reprogramming efficiencies, kinetics and tumorigenicity vary between somatic cell types. Firstly, cell sources have to be permissive to avoid to turn to integrative methods and to the use of oncogenes. Some human, adult somatic cells, such as melanocytes, are known to naturally express endogenously reprogramming factors, for instance Sox 2, at sufficiently high levels [18, 19]. Moreover, some types of donor cells such as dermal fibroblasts and blood cells are easily accessible, but they might carry more mutational burdens and chromosomal abnormalities, due to their frequent exposure to environmental stress factors, like ultraviolet rays, or due to the donor's age, thereby leading to increased tumorigenicity, and significant safety problems [20, 21]. With all these considerations of cell variability and tumorigenic potential in mind, reflection on the generation of homogeneous cell source and banking emerged.

Many approaches have been evaluated to address the tumorigenicity challenge by eliminating the pluripotent cells of the final product such as small molecule, genetic approach to introduce a suicide gene; miRNA switch; antibodies targeting a surface-specific antigen; phototoxic approach; live detection and quantification of the residual human iPSC [22]. For the suicide gene approach, the most widely used gene is herpes simplex virus thymidine kinase (HSV-TK) that phosphorylates ganciclovir (GCV) and induces apoptosis by inhibiting DNA synthesis. Many studies demonstrated its efficacy as safeguard to eliminate tumoral cells [23]. Until then, this genetic approach with an inducible suicide system may remain not necessary enough to induce tumor elimination because of potential acquired resistance to GCV due to variability of insertion location sites and to the uncontrolled number of inserted transgene [24]. Another study demonstrated the same mechanism of inducing apoptosis in 95% of iPSCs and iPSC-derived cells by transducing an inducible Caspase 9 [25]. Recently, with development of targeted genetic strategies such as gene-editing, researchers try to identify the location of "genomic safe harbors" (GSH), corresponding to the safest permissive loci for transgenes' insertion [26]. The already known GSH candidates could be AAVS1 (adeno-associated virus integration site 1), CCR5 (chemokine CC motif receptor 5), human ROSA26 and some extragenic loci. Recently, to predict the influence of gene integration on nearby genes, it has been suggested that the combination of several distinct approaches such as the analysis of the topologically associated domains of GSH candidates of chromosomes could reduce the risks associated with cell therapy [27]. Another targeted alternative, eliminating selectively residual pluripotent cells sparing precursors and differentiated cells, involves PluriSIns, pluripotent cells-specific inhibitors [28]. Alternatively, antibody, lectin or miRNA-mediated removal undesired cells were developed to suppress the pluripotent stem cells from the final product [29]. Lastly, a novel methodology using synthetic microRNA switch is developed to improve the purity of the final product even if the cell surface markers are not available to tag the relevant cells [30, 31].

3.2 iPSC immunogenicity

The immunogenicity of differentiated cells derived from iPSC is of clinical significance. At the beginning, because of the use of the patient's own cells, theoretically there is no risk of rejection after their transplantation. Some studies demonstrated no immune rejection of autologous iPSC-derived cells, but an activated immune response after the use of allogeneic iPS derived cells. Contrarily, immune rejection has been observed after autologous transplantation of iPSC-derived cells, suggesting that *in vitro* operations could also impact on the immunogenicity of the iPSC [32]. Moreover, the immune response to undifferentiated iPSC is different from their derivatives, emphasizing the need to perform similar comparative analyses in starting cell populations in order to predict immune tolerance after transplantation. Whereas autologous hiPSC-derived smooth muscle cells were highly immunogenic, autologous hiPSC-derived retinal pigment epithelial (RPE) cells were immune tolerated, suggesting a potential abnormal expression of some immunogenic antigens in smooth muscle cells [33]. These results demonstrated that the nature of the differentiated cells could trigger an immune response suggesting the importance of the differentiation protocol.

As mentioned earlier, because of their genomic instability, generation, amplification and differentiation of iPSC could induce a modified immune response of the iPSC *in vivo*. Concerning reprogramming, the RNA-based methods are relatively efficient and do not integrate in the genome, but they are also known to be highly immunogenic. Concerning cell type, it has been widely shown that iPSCs could be generated from a patient's own cells including fat cells, nerve cells, skin fibroblasts, cuticle cells, fetal foreskin cells, B cells, T cells, peripheral blood mononuclear cells, umbilical cord mesenchymal cells, chorionic mesenchymal cells and amniotic mesenchymal cells. But, some studies showed that the genetic memory of the cellular immunogenicity is conserved after reprogramming and differentiation. So, the selection of donor cell type/origin is crucial. As an example, iPSCs derived from less immunogenic cells, such as umbilical cord mesenchymal cells, generated less immunogenic neural derivatives than those from skin fibroblasts-derived iPSCs [34]. Recently, several researchers showed the less immunogenic potential of some iPSC-derived cells as cartilage and retinal pigment epithelium cells when they are implanted *in vivo*, arguing that some cell types are less immunogenic and should be preferred for clinical settings [35, 36].

Recently, a novel approach of "Universal" iPSC was developed to address the difficulty of immunogenicity of allogeneic iPSCs. Hypoimmunogenicity of iPSC was induced by inactivation of major histocompatibility complex class I and II genes and overexpression of CD47 enabled them to escape to immune rejection in fully HLA-mismatched allogeneic recipients. This strategy allowed the long-term survival of the transplanted cells without the use of immunosuppression. However, overexpression of CD47 is associated with malignant transformation, leading to include some suicide strategies as a safety concern [37]. These immune escape approaches open the door to the clinical use of allogeneic iPSC-derived cell products without immune rejection concerns and complications. However, their complex production process including a combination of several transduction and gene-editing operations could add many safety issues. Even though other vectors and gene-editing techniques [38, 39] could also be used to reduce the risks, the multiple genetic manipulations and additional expansions in culture require a reinforced control of the "Universal" iPSC quality for clinical settings.

4. iPSC for clinical use

4.1 Clinical-grade allogeneic iPSC line bank

The use of human iPSCs in medicinal applications requires the establishment of standardized and validated protocols that will allow large-scale, cost-effective cultivation procedure, while maintaining their quality. Implementation of good manufacturing practice (GMP)-compliant protocols for the generation and maintenance of human iPSC lines is crucial to increase the application safety and to fulfill the regulatory requirements to obtain clinical trials' approval. Many efforts to increase the overall iPSC stability, reproducibility and quality have been performed by (1) selecting the cell type that is easily accessible, less immunogenic, and permissive for reprogramming and presents the ability to be stored for longer periods of time; (2) improving reprogramming efficiency, which should be as high as possible without genomic integration-based delivery method and without using oncogene and (3) improving cultivation methods with xeno- and feeder-free products, with defined and scalable conditions for maintenance and differentiation of human iPSC such as automation, closed cell systems and validated protocols [40]. Moreover, selection of cell source is of importance. Demonstration of comparability, standardization and validation of such systems is critical for iPSC-derived therapies. To circumvent and manage the safety risk of the iPSC for regenerative medicine, several groups worked at the early stage on the development of standardized clinical grade iPSC banks from allogeneic donors. Indeed, the use of highly defined iPSC as starting cells presents many advantages as overcoming the genetic variations inducing different immunogenicity, genetic instability, tumorigenicity, and differentiation outcomes. Moreover, generation of iPSC from each patient is costly and time-consuming. In this regard, several groups in the world have developed banking of allogeneic iPSC lines for clinical use with validated and standardized protocols. The possibility of creating off-the-shelf iPSC-based therapies has attracted not only academics but also industrial groups as Lonzo and Cellular Dynamics International, a Fujifilm company.

iPSC banks can provide a cost-effective mass-production strategy. Several groups have developed iPSC banks from selected HLA donors trying to cover the majority of the population [41, 42]. The Center for iPSC Research and Application (CiRA), in Kyoto University, started the iPS Cell Stock for Regenerative Medicine in 2013. Initially, based on the limited diversity of the Japanese population, CiRA wanted to generate clinical-grade iPSCs from samples of peripheral blood and umbilical cord blood from healthy selected donors that would cover 90% of Japanese population with only 50 iPSC lines [43]. This strategy is valuable for countries such as Japan, but could be difficult to expand to the worldwide population. It has been evaluated that a multiethnic iPSC bank of the 100 most common HLA types in each population would cover only 78% of European individuals, 63% of Asians, 52% of Hispanics and 45% of African Americans [44]. This probabilistic model highlights the necessity of a large-scale international collaboration for the constitution of haplobank of iPSC lines. Using HLA-homozygous donors limits the numbers of iPSC lines needed to cover a given population, but identification of the potential donors would need large screenings or the use of established data from cord blood banks. The potential development of "universal" iPSCs made of genetically modified cells offering an off-the-shelf product that is readily available could be an alternative to the iPSC bank using materials from HLA-homozygous donors. The "universal" iPSC could solve the problem of immune rejection profile of iPSC-derived cells by artificially expressing, for example, HLA molecule as HLA-E allowing iPSC-derived cells to escape T cell-mediated rejection and to be resistant to NK-cell lysis [37, 45].

Nevertheless, stochastic events potentially occurring during reprogramming, colony expansion, iPSC selection, differentiation, iPSC-derived cell expansion and purification, storage and transport could complicate efforts toward a standardized product. Consequently, it has to be taken into consideration that variation may exist within any iPSC bank, between iPSC and final product composed of iPSC-derived cells in the clinic. Such variability requires continual extensive genotypic, phenotypic and functional assessment and highlights the need of a global quality control confirming the iPSC and the iPSC-derived cells' quality whatever the manufacturer, the reprogramming method or the cell donors.

4.2 Quality control of clinical-based iPSC

Given the high variability across iPSC lines and their differentiated derivatives in terms of their epigenetic status, tumorigenic and immunogenic potential, differentiation capacity, batch variability and existence of heterogeneous populations and/or non-relevant cells such as contaminating cell, the clinical outcome of the cell replacement therapy, in terms of efficacy and safety with these iPSC-based products, highly relies on the acceptable quality and safety standards of these products. Because of dissimilarities between institutions on these criteria, agreement on the critical quality attributes (CQAs) of such lines and the assays that should be used is required. The CQAs correspond to the chemical, physical and biological properties of the product. As well as the type of assay, they have to be defined within an appropriate limit, range or distribution to ensure quality and safety of the product. For cell therapy product and for clinical-grade iPSC, the CQAs include identity, microbiological sterility, genetic fidelity and stability, viability, characterization and potency. In the last few years, there was a common effort made on the banking and the quality control of the iPSC lines. After a series of workshop, adaptation to iPSC of the established recommendations and guidance realized by the International Stem Cell Banking Initiative (ISCBI) for human embryonic stem cell banking, has generated initial recommendations on the minimum dataset required to consider an iPSC line of clinical grade [46]. During these workshops, the researchers, industrial and regulation agencies pointed out the requirement of standardization and validation of process and quality and safety controls. For each criterion, one or several tests are required with regard to the recommended analytical methods. Global consensus recommends the performance of assays by accredited and licensed laboratories. When it is not available, in-house tests should be undertaken after validation and qualification, and comparability with other laboratories should be performed if possible.

The first mandatory test is to validate the identity of the iPSC line with the short tandem repeat (STR) analysis to genotype the original cells, the iPSC seeds and the master cell bank to ascertain the absence of switch or cross contamination of several iPSC lines during generation or maintenance process. Due to the nature of the stem cell-based products, they cannot be sterilized. The assessment of the microbiological sterility is of the highest importance and should be performed not only on the final product. This should include the mycoplasma, bacteriology and viral testing supplemented by endotoxins detection assay and should have a negative result. The genetic stability and fidelity of the iPSC lines should be evaluated by residual vector testing and karyotype. To eliminate the risk of potential cell transformation and the risk of malignancy development in patients, residual vector testing has to be ≤1 plasmid copy per 100 cells in seed and master cell banks and the karyotype should be normal on more than 20 metaphases. So far, techniques with high precision such as single nucleotide polymorphism (SNP) and whole genome analysis or other genetic markers are not required but could be performed for information. To give an appropriate dosage of cells, viability should be >60%. Calculation of doubling time and detection of cell debris are not required but could provide useful information. To manage the risk associated with the presence of non-desired or spontaneously differentiated cells, iPSCs have to be characterized by the expression of a minimum of two markers from the standard human pluripotent stem cells panel (positive for Oct4, TRA-1-60, TRA-1-81, SSEA-3, SSEA-4, Sox2, Nanog). A combination of one intracellular and one extracellular marker should be used and should be >70%. Finally, for the potency assay, reflecting the biological activity of the cells, embryoid body formation or directed differentiation of monolayer cultures to produce cell types representative of all three embryonic germ layers is mandatory. The teratoma formation in severe combined immune-deficient (SCiD) mouse injection assay is not mandatory for the iPSC due to a reproducibility problem, high cost and non-ethical procedure. Molecular pluripotency assays such as mRNA array- and RNA-Seq-based gene expression assays could be kept for information if they are performed molecular pluripotency assays such as mRNA array- and RNA-Seq-based gene expression assays could be for information but are not required. For the iPS-derived differentiated therapeutic products, the minimal criteria are mostly identical except for the phenotypical characterization, which should validate the absence of pluripotent stem cell markers, the expression of differentiation markers unique to the therapeutic product and assess 100% purity of the therapeutic cellular product without any contaminating other lineage cell types.

This consensus on CQA and minimum testing requirements for clinical-grade iPSC lines will evolve with the advances in scientific understanding and development in technology and best practices. The Global Alliance for iPSC Therapies (GAiT), which facilitates the development of general clinical-grade iPSC standards by community engagement and consensus building to support the global application of iPSC-derived cellular therapeutics, is in charge of the future evolution of the consensus on quality and safety standards required for a clinical-grade iPSC. Moreover, GAiT presents objectives to achieve consensus on donor selection and screening criteria and consent standards, which with future commercialization and global distribution also require ethical review.

5. Conclusion

It is quite remarkable that in just over 10 years, research using iPSC has led to several clinical studies, with many more applications expected to follow. In few years, the iPSC-based therapies induced a switch to a mass production of clinical-grade iPSC for the benefit of a large population at affordable costs, with the generation of clinical-grade iPSC banks, and with a stronger involvement of biopharmaceutical companies. This shift led to many efforts for the standardization of generation, maintenance and differentiation procedures, and for the establishment of quality and safety standards for the clinical-grade iPSC and their derivatives prior to transplantation to patients.

There are still a number of challenges that must be overcome for iPSCs to reach their full potential. The improvement of manufacturing procedures for a large-scale production would provide higher quality cells for clinical iPSC-based therapies. Quality and safety controls are also challenging. Predicting cancer risk based on sequence information is a formidable task, and failure to detect oncogenic mutations is not necessarily a warrantor of the non-tumorigenicity of iPSC-based products, suggesting that recommendations should still evolve with scientific advances.

Due to their large potential in regenerative medicine, such as the generation of complex 3D structures, tissues or organs, more challenges in differentiation protocols in 3D structures have to be overcome for the up-coming year, without compromising quality and safety of iPSCs.

Conflict of interest

The authors declare no conflict of interest.

Author details

Valérie Vanneaux
Department of Biotherapies, Saint-Louis Hospital, Paris, France

*Address all correspondence to: valerie.vanneaux@aphp.fr

References

[1] Takahashi K, Yamanaka S. Induction of pluripotent stem cells from mouse embryonic and adult fibroblast cultures by defined factors. Cell. 2006;**126**:663-676. DOI: 10.1016/j.cell.2006.07.024

[2] Eguizabal C, Aran B, Chuva de Sousa Lopes SM, Geens M, Heindryckx B, Panula S, et al. Two decades of embryonic stem cells: A historical overview. Human Reproduction Open. 2019;**2019**(1):1-17. DOI: 10.1093/hropen/hoy024. eCollection 2019

[3] Mandai M, Watanabe A, Kurimoto Y, Hirami Y, Morinaga C, Daimon T, et al. Autologous induced stem-cell-derived retinal cells for macular degeneration. The New England Journal of Medicine. 2017;**376**:1038-1046. DOI: 10.1056/NEJMc1706274

[4] Cynata Therapeutics [Internet]. 2019. Available from: https://www.cynata.com/graftversushostdiseaseaftversushost disease [Accessed: 30 June 2019]

[5] Fate Therapeutics [Internet]. 2019. Available from: https://fatetherapeutics.com/pipeline/immuno-oncology-candidates/ft500 [Accessed: 30 June 2019]

[6] Rossignol J, Crane AT, Fink KD, Dunbar GL. Will undifferentiated induced pluripotent stem cells ever have clinical utility? Journal of Stem Cell Research and Therapy. 2014;**4**:189

[7] Galat V, Galat Y, Perepitchka M, Jennings LJ, Iannaccone PM, Hendrix MJ. Transgene reactivation in induced pluripotent stem cell derivatives and reversion to pluripotency of induced pluripotent stem cell-derived mesenchymal stem cells. Stem Cells and Development. 2016;**25**(14):1060-1072. DOI: 10.1089/scd.2015.0366

[8] Nori S, Okada Y, Nishimura S, Sasaki T, Itakura G, Kobayashi Y, et al. Long-term safety issues of iPSC-based cell therapy in a spinal cord injury model: Oncogenic transformation with epithelial-mesenchymal transition. Stem Cell Reports. 2015;**4**:360-373. DOI: 10.1016/j.stemcr.2015.01.006

[9] Griscelli F, Desterke C, Feraud O, Divers D, Oudrhiri N, Tosca L, et al. Genomic landscape analyses of reprogrammed cells using integrative and non-integrative methods reveal variable cancer-associated alterations. Oncotarget. 2019;**10**(28):2693-2708. DOI: 10.18632/oncotarget.26857

[10] Feng C, Jia YD, Zhao XY. Pluripotency of induced pluripotent stem cells. Genomics, Proteomics & Bioinformatics. 2013;**11**(5):299-303. DOI: 10.1016/j.gpb.2013.08.003

[11] Melton C, Judson R, Blelloch R. Opposing microRNA families regulate self-renewal in mouse embryonic stem cells. Nature. 2010;**463**(7281):621-626. DOI: 10.1038/nature08725

[12] Zhou H, Wu S, Joo JY, Zhu S, Han DW, Lin T, et al. Generation of induced pluripotent stem cells using recombinant proteins. Cell Stem Cell. 2009;**4**:381-384. DOI: 10.1016/j.stem.2009.04.005

[13] Liang G, Zhang Y. Genetic and epigenetic variations in iPSCs: Potential causes and implications for application. Cell Stem Cell. 2013;**13**(2):149-159. DOI: 10.1016/j.stem.2013.07.001

[14] Kim K, Doi A, Wen B, Ng K, Zhao R, Cahan P, et al. Epigenetic memory in induced pluripotent stem cells. Nature. 2010;**467**:285-290. DOI: 10.1038/nature09342

[15] Polo JM, Liu S, Figueroa ME, Kulalert W, Eminli S, Tan KY, et al. Cell

type of origin influences the molecular and functional properties of mouse induced pluripotent stem cells. Nature Biotechnology. 2010;**28**:848-855. DOI: 10.1038/nbt.1667

[16] Ghosh Z, Wilson KD, Wu Y, Hu S, Quertermous T, Wu JC. Persistent donor cell gene expression among human induced pluripotent stem cells contributes to differences with human embryonic stem cells. PLoS ONE. 2010;**5**:e8975. DOI: 10.1371/journal.pone.0008975

[17] Boland MJ, Nazor KL, Loring JF. Epigenetic regulation of pluripotency and differentiation. Circulation Research. 2014;**115**:311-324. DOI: 10.1161/CIRCRESAHA.115.301517

[18] Utikal J, Maherali N, Kulalert W, Hochedlinger K. Sox2 is dispensable for the reprogramming of melanocytes and melanoma cells into induced pluripotent stem cells. Journal of Cell Science. 2009;**122**:3502-3510. DOI: 10.1242/jcs.054783

[19] Kim JB, Greber B, Arauzo-Bravo MJ, Meyer J, Park KI, Zaehres H, et al. Direct reprogramming of human neural stem cells by OCT4. Nature. 2009;**461**:649-643. DOI: 10.1038/nprot.2009.173

[20] Lo Sardo V, Ferguson W, Erikson GA, Topol EJ, Baldwin KK, Torkamani A. Influence of donor age on induced pluripotent stem cells. Nature Biotechnology. 2017;**35**(1):69-74. DOI: 10.1038/nbt.3749

[21] Musunuru K, Sheikh F, Gupta RM, Houser SR, Maher KO, Milan DJ, et al. Induced pluripotent stem cells for cardiovascular disease modeling and precision medicine: A scientific statement from the American Heart Association. Circulation: Genomic and Precision Medicine. 2018;**11**(1):1-30. DOI: 10.1161/HCG.0000000000000043

[22] Jeong HC, Cho SJ, Lee MO, Cha HJ. Technical approaches to induce selective cell death of pluripotent stem cells. Cellular and Molecular Life Sciences. 2017;**74**(14):2601-2611. DOI: 10.1007/s00018-017-2486-0

[23] Schuldiner M, Itskovitz-Eldor J, Benvenisty N. Selective ablation of human embryonic stem cells expressing a "suicide" gene. Stem Cells. 2003;**21**:257-265. DOI: 10.1634/stemcells.21-3-257

[24] Kotini AG, de Stanchina E, Themeli M, Sadelain M, Papapetrou EP. Escape mutations, ganciclovir resistance, and teratoma formation in human iPSCs expressing an HSVtk suicide gene. Molecular Therapy—Nucleic Acids. 2016;**5**:e284. DOI: 10.1038/mtna.2015.57

[25] Itakura G, Kawabata S, Ando M, Nishiyama Y, Sugai K, Ozaki M, et al. Fail-safe system against potential tumorigenicity after transplantation of iPSC derivatives. Stem Cell Reports. 2017;**8**:673-684. DOI: 10.1016/j.stemcr.2017.02.003

[26] Sadelain M, Papapetrou EP, Bushman FD. Safe harbours for the integration of new DNA in the human genome. Nature Reviews. Cancer. 2012;**12**:51-58. DOI: 10.1038/nrc3179

[27] Kinamura Y, Shofuda T, Higuchi Y, Nagamori I, Oda M, Nakamori M, et al. Human genomic safe harbors and the suicide gene-based safeguard system for iPSC-based cell therapy. Stem Cells Translational Medicine. 2019;**8**(7):627-638. DOI: 10.1002/sctm.18-0039

[28] Ben-David U, Biran A, Scaffidi P, Herold-Mende C, Boehringer M, Meshorer E, et al. Elimination of undifferentiated cancer cells by pluripotent stem cell inhibitors. Journal of Molecular Cell Biology. 2014;**6**:267-269. DOI: 10.1093/jmcb/mju012

[29] Lynch CL, Altun G, Tran HT, Garitaonandia I, Slavin I, Loring JF, et al. Specific lectin biomarkers for isolation of human pluripotent stem cells identified through array-based glycomic analysis. Cell Research. 2011;**21**:1551. DOI: 10.1038/cr.2011.148

[30] Miki K, Endo K, Takahashi S, Funakoshi S, Takei I, Katayama S, et al. Efficient detection and purification of cell populations using synthetic microRNA switches. Cell Stem Cell. 2015;**16**:699-711. DOI: 10.1016/j.stem.2015.04.005

[31] Parr CJC, Katayama S, Miki K, Kuang Y, Yoshida Y, Morizane A, et al. MicroRNA-302 switch to identify and eliminate undifferentiated human pluripotent stem cells. Scientific Reports. 2016;**6**:32532. DOI: 10.1038/srep32532

[32] Zhao T, Zhang ZN, Rong Z, Xu Y. Immunogenicity of induced pluripotent stem cells. Nature. 2011;474(7350):212-215. DOI: 10.1038/nature10135

[33] Zhao T, Zhang ZN, Westenskow PD, Todorova D, Hu Z, Lin T, et al. Humanized mice reveal differential immunogenicity of cells derived from autologous induced pluripotent stem cells. Cell Stem Cell. 2015;**17**(3):353-359. DOI: 10.1016/j.stem.2015.07.021

[34] Liu S, Yuan M, Hou K, Zhang L, Zheng X, Zhao SX, et al. Immune characterization of mesenchymal stem cells in human umbilical cord Wharton's jelly and derived cartilage cells. Cellular Immunology. 2012;**278**(1-2):35-44. DOI: 10.1016/j.cellimm.2012.06.010

[35] Nguyen D, Hägg DA, Forsman A, Ekholm J, Nimkingratana P, Brantsing C, et al. Cartilage tissue engineering by the 3D bioprinting of iPS cells in a nanocellulose/alginate Bioink. Scientific Reports. 2017;7(1):658. DOI: 10.1038/s41598-017-00690-y

[36] Westenskow PD, Bucher F, Bravo S, Kurihara T, Feitelberg D, Paris LP, et al. iPSC-derived retinal pigment epithelium allografts do not elicit detrimental effects in rats: A follow-up study. Stem Cells International. 2016;**2016**:8470263. DOI: 10.1155/2016/8470263

[37] Deuse T, Hu X, Gravina A, Wang D, Tediashvili G, De C, et al. Hypoimmunogenic derivatives of induced pluripotent stem cells evade immune rejection in fully immunocompetent allogeneic recipients. Nature Biotechnology. 2019;**37**:252-258. DOI: 10.1038/s41587-019-0016-3

[38] Roy KR, Smith JD, Vonesch SC, Lin G, Tu CS, Lederer AR, et al. Multiplexed precision genome editing with trackable genomic barcodes in yeast. Nature Biotechnology. 2018;**36**:512-520. DOI: 10.1038/nbt.4137

[39] Strohkendl I, Saifuddin FA, Rybarski JR, Finkelstein IJ, Russell R. Kinetic basis for DNA target specificity of CRISPR-Cas12a. Molecular Cell. 2018;**71**:816-824.e3. DOI: 10.1016/j.molcel.2018.06.043

[40] Dakhore S, Nayer B, Hasegawa K. Human pluripotent stem cell culture: Current status, challenges, and advancement. Stem Cells International. 2018;**2018**:7396905. DOI: 10.1155/2018/7396905

[41] Okita K. iPS cells for transplantation. Current Opinion in Organ Transplantation. 2011;**16**(1):96-100. DOI: 10.1097/MOT.0b013e32834252a2

[42] Taylor CJ, Peacock S, Chaudhry AN, Bradley JA, Bolton EM. Generating an iPSC bank for HLA-matched tissue transplantation based on known donor and recipient HLA types. Cell Stem Cell. 2012;**11**(2):147-152. DOI: 10.1016/j.stem.2012.07.014

[43] Tokunaga K, Nakatsuji N, Nakajima F. HLA-haplotype banking and iPS cells. Nature Biotechnology. 2008;**26**(7):739-740. DOI: 10.1038/nbt0708-739

[44] Gourraud P-A, Gilson L, Girard M, Peschanski M. The role of human leukocyte antigen matching in the development of multiethnic "haplobank" of induced pluripotent stem cell lines. Stem Cells. 2012;**30**(2):180-186. DOI: 10.1002/stem.772

[45] Gornalusse GG, Hirata RK, Funk SE. HLA-E-expressing pluripotent stem cells escape allogeneic responses and lysis by NK cells. Nature Biotechnology. 2017;**35**:765-772. DOI: 10.1038/nbt.3860

[46] Sullivan S, Stacey GN, Akazawa C, Aoyama N, Baptista R, Bedford P, et al. Quality control guidelines for clinical-grade human induced pluripotent stem cell lines. Regenerative Medicine. 2018;**13**:859-866. DOI: 10.2217/rme-2018-0095

Section 2

Clinical Utilization of Mesenchymal Stem Cells

Chapter 3

The Rising Role of Mesenchymal Stem Cells in the Treatment of Various Infectious Complications

Khalid Ahmed Al-Anazi, Waleed K. Al-Anazi and Asma M. Al-Jasser

Abstract

Mesenchymal stem cells are heterogenous adult multipotent stromal cells that can be isolated from various sources including: bone marrow, peripheral blood, umbilical cord blood, dental pulp, and adipose tissue. They have certain immunomodulatory, immunosuppressive, and antimicrobial properties that enable them to have several therapeutic and clinical applications including: treatment of autoimmune disorders, role in hematopoietic stem cell transplantation and regenerative medicine, as well as treatment of various infections and their associated complications such as septic shock and acute respiratory distress syndrome. Although more success has been achieved in preclinical trials on the use of mesenchymal stem cells in animal models than in human clinical trials, particularly in septic shock and Chagas disease, more progress has been made in both disorders after the recent use of specific sources and certain doses of mesenchymal stem cells. Nevertheless, the utilization of this type of stem cells has shown remarkable progress in the treatment of few infections such as tuberculosis. The clinical application of mesenchymal stem cells in the treatment of several diseases still faces real challenges that need to be resolved. The following book chapter will be an updated review on the role of mesenchymal stem cells in various infections and their complications.

Keywords: mesenchymal stem cells, host immunity, antimicrobial properties, septic shock, *Mycobacterium tuberculosis*, Chagas disease, human immunodeficiency virus

1. Introduction to mesenchymal stem cells

Mesenchymal stem cells (MSCs), which were first described by Alexander Fridenstein in the 1960s, are heterogeneous, non-hematopoietic, adult multipotent stromal progenitor cells that are capable of self-renewal as well as differentiation into multiple lineages and various cell types [1–8]. They can be isolated from several sources including bone marrow (BM), peripheral blood (PB), umbilical cord blood (UCB), amniotic fluid, placenta, adipose tissue (AT), and dental pulp as shown in **Table 1** [1–8]. Although the BM is the main source of MSCs, these stromal cells constitute only a small fraction of the total number of cells populating the BM [2, 4–6].

MSCs have the following distinguishing features: (1) ability to adhere to the plastic vessel under optimal culture conditions; (2) capability to differentiate into osteoblasts, adipocytes, and chondrocytes; and (3) having characteristic immunophenotypic profile on flow cytometry [1–3, 5, 6, 8, 9]. MSCs are characteristically positive for: CD 105, CD 73, and CD 90 and characteristically negative for the following surface

1	Bone marrow
2	Peripheral blood
3	Umbilical cord blood: Wharton's jelly
4	Placenta: chorionic villi of placenta
5	Amniotic fluid
6	Menstrual blood
7	Fallopian tubes and cervical tissue
8	Breast milk
9	Adipose tissues: fat
10	Dental pulp, periodontal ligaments, and exfoliated deciduous teeth
11	Palatal tonsils
12	Salivary glands
13	Skeletal muscle tissues
14	Dermal tissues
15	Lung tissues and alveolar epithelium
16	Liver tissues: fetal liver
17	Synovial membrane and fluid
18	Parathyroid glands

Table 1.
Sources of mesenchymal stem cells.

	Positive		Negative
Characteristic surface markers	**CD 105** **CD 73** **CD 90**		**CD 45** **CD 34** **CD 14** **CD 11b** **CD 19** **CD 79a** **HLA-DR**
Other surface markers that may/may not be expressed	CD 117 CD 166 CD 29 CD 44 CD 106 CD 9 CD 10 CD 13 CD 28	CD 33 CD 49b CD 71 CD 164 CD 271 HLA-class I Stro-1 SSEA-4 ITGA-11	CD 31 CD 33 CD 133

MSCs, mesenchymal stem cells; HLA, human leukocyte antigen.
The bold values are to differentiate characteristic from non-characteristic surface markers.

Table 2.
Surface markers of MSCs on Flow cytometry.

markers: CD 45, CD 34, CD 11b, CD 14, CD 19, CD 79a, and HLA-DR. However, certain types of MSCs can occasionally show positivity or negativity for specific surface markers as shown in **Table 2** [1–3, 5, 6, 8–14]. Also, MSCs can differentiate into other cell types including: myocytes, cardiomyocytes, and neurons [5].

Several studies have shown that MSCs obtained from BM, AT, and other sources do express CD 34 surface markers [4, 15–18]. MSCs can be seen in abundant numbers in the circulation under the following circumstances: stem cell mobilization with growth factors, tissue injuries, stroke, hypoxia, and inflammatory conditions [4, 19–24]. Despite the efforts displayed over the last five decades including identification of nine transcriptional factors, little is known about the molecular basis underlying the stemness of MSCs and it is still unclear whether these recently discovered genes regulate stemness or only differentiation of MSCs [7].

2. Functions, properties, and therapeutic indications of MSCs

MSCs have immunomodulatory and immunosuppressive properties that enable them to have several therapeutic and clinical applications including: hematopoietic stem cell transplantation (HSCT), autoimmune disorders, regenerative medicine and tissue repair, neurological diseases, bone and cartilage disorders, as well as treatment of several infections and acute respiratory distress syndrome (ARDS). Details are shown in **Table 3** [1, 2, 6, 8, 25–29]. MSCs are major constituents of the BM microenvironment and the HSC niche and apparently they are the masters of

1. Hematopoietic stem cell transplantation:
    a. Enhancement of engraftment
    b. Prevention of graft versus host disease (GVHD)
    c. Treatment of GVHD
2. Treatment of autoimmune diseases:
    a. Systemic lupus erythromatosus
    b. Rheumatoid arthritis
    c. Systemic sclerosis
    d. Type 1 diabetes mellitus
    e. Multiple sclerosis
    f. Crohn's disease
3. Regenerative medicine and tissue repair:
    a. Myocardial ischemia
    b. Cardiac dysfunction
    c. Chronic non-healing wounds
    d. Liver injury
    e. Myocardial infarction
    f. Dilated cardiomyopathy
    g. Critical limb ischemia
    h. Spinal cord injuries
4. Treatment of various infections:
    a. Bacterial infections including sepsis and its associated acute respiratory distress syndrome
    b. Viral infections such as human immunodeficiency virus, hepatitis B and C viruses
    c. Parasitic infections such as Chagas disease, schistosomiasis, and malaria
    d. Mycobacterial infections such as tuberculosis
5. Other indications:
    a. Macular degeneration, corneal reconstruction and transplantation
    b. Bones and joints: osteogenesis imperfecta, osteoarthritis, and osteoporosis
    c. Cancer gene therapy
    d. Amyotrophic lateral sclerosis
    e. Liver cirrhosis

Table 3.
Current and potential therapeutic indications for mesenchymal stem cells.

survival and clonality [30–32]. The main functions of MSCs include: formation of hematopoietic microenvironment, modulation of the activity of the immune system, and regulating cell trafficking [33].

3. Role of MSCs in host defense and infections

The putative roles of BM-MSCs during infection are: (1) detection of pathogens, (2) activation of host immune responses, (3) elimination of pathogens, (4) induction of proinflammatory gradients, and (5) modulation of proinflammatory host immune response due to having specific immunoregulatory properties of MSCs including: inhibition of differentiation of monocytes to dendritic cells (DCs), alteration of cytokine profile of DCs, induction of tolerant phenotypes of naïve and effector T-cells, inhibition of antibody production by B-cells, and suppression of natural killer (NK) cell proliferation and NK-mediated cytotoxicity [1, 2, 28, 34]. BM-MSCs may augment antimicrobial responses, abridge proinflammatory and damage responses, and ameliorate associated tissue injury and they appear to function as a critical fulcrum providing balance by promoting pathogen clearance during the initial inflammatory response, and suppressing inflammation to preserve integrity of the host and facilitate tissue repair [1, 2, 34].

The immunomodulatory properties of MSCs are mediated by cell-to-cell interaction and the secreted cytokines [35–37]. MSCs could potentially be involved at multiple levels in host defense by mobilizing immune effector cells and modulation of proinflammatory immune responses to minimize tissue damage [1, 37]. BM-MSCs may protect against infectious challenge by direct effects on the pathogens or through indirect effects on the host [1]. However, placenta-derived MSCs and fetal membrane-derived MSCs are highly susceptible to herpes viruses including varicella zoster virus (VZV) [2, 38]. Several types of stem cells including BM-MSCs and neural stem cells can cross the blood brain barrier and reach not only brain tumors but also ischemic and injured tissues caused by certain infections in the brain and engraft there. Consequently, MSCs can be utilized as means of cellular carriers to deliver therapeutic agents to sites of brain injury in order to exert their therapeutic and tissue regenerative effects in the brain [39–43].

4. Antimicrobial properties of MSCs

MSCs have been shown to exhibit the following antimicrobial properties: (1) capacity to enhance antibacterial activity by interaction with the host innate immune system in order to increase antibiotic sensitivity, increase bacterial killing, and slow bacterial growth; (2) capacity to enhance bacterial clearance in preclinical models of sepsis, cystic fibrosis, and ARDS; and (3) secretion of antimicrobial peptides such as: interleukin (IL)-17, indoleamine 2,3-dioxygenase (IDO), β-defensins, lipocalin-2, and cathelicidin LL-37 [44–46]. Members of the chemokine family have been found to have antimicrobial peptide activity although the role of chemokines in immunity during infection is rather complicated [47].

5. MSCs in sepsis, ARDS, and chronic bacterial infections

5.1 MSCs in sepsis syndrome and septic shock

Sepsis syndrome and septic shock represent major health problems worldwide and they are leading causes of death in hospitalized patients due to their association with high rates of morbidity and mortality in the absence of effective therapy [48–51]. Sepsis is a potentially lethal syndrome that can develop following an infection in which a breakdown in the immune homeostasis results in both proinflammatory and anti-inflammatory mechanisms that become uncoupled from normal regulation [50]. The inflammatory-driven maladaptive response induces disruption of endothelial and epithelial barriers, thus resulting in organ dysfunction. However, the host responds to sepsis by stimulating the proliferation of HSCs in the BM or by activating emergency hematopoiesis in an attempt to counteract the effects of sepsis on the function of multiple body organs [51]. Septic shock is a devastating complication of uncontrolled bacterial infection that carries a mortality rate of 20–50% [50, 52]. Currently, there is no specific treatment for septic shock and the management of this devastating complication of serious infections remains supportive. However, the following measures should be taken into consideration: early identification, fluid resuscitation, prompt institution of antibiotic therapy, control of the source of infection, circulatory support, and lung protection by mechanical ventilation [48, 49, 52, 53].

Based on numerous preclinical studies, cell-based therapies are potentially beneficial in the treatment of septic shock and ARDS. However, various types of stem cells including embryonic stem cells, MSCs, and induced pluripotent stem cells have been used in the treatment of sepsis and ARDS, but MSCs are the most commonly used stem cells in septic shock [53]. In patients with septic shock complicated by acute lung injury (ALI) and ARDS, the paracrine factors secreted by MSCs can: mediate endothelial and epithelial permeability, and increase alveolar fluid clearance in addition to other mechanisms that reduce the complications of septic shock [54].

In a mouse model of sepsis, lipopolysaccharide-preconditioned MSC transplantation has been shown to: ameliorate survival rate after transplantation, protect cells from apoptosis and organ damage, and have immunomodulatory therapeutic properties [55]. Also, transplanted MSC can secrete Toll-like receptor-4, which plays a seminal role in attenuating in vivo *Escherichia coli*-induced pneumonia and ALI through anti-inflammatory and antibacterial effects [56]. In experimental animal models of sepsis, the effectiveness of BM-MSCs was compared to that of Wharton's jelly (WJ) of umbilical cord; both sources of MSCs regulated leukocyte trafficking and reduced organ dysfunction but only WJ-MSCs were able to improve bacterial clearance and survival [57]. In animal models of Staphylococcal toxic shock syndrome, MSCs; particularly AT derived MSCs; were able to suppress cytokine production and attenuate sepsis but they failed to improve survival [58, 59].

Several preclinical sepsis studies have suggested that MSCs are capable of: modulating inflammation, enhancing clearance of pathogens as well as tissue repair, thus resulting in improvement in symptoms and reduction in organ damage and finally improvement in survival and reduction in mortality rates [48–50, 52]. A meta-analysis that evaluated the preclinical use of MSCs in animal models of septic shock demonstrated that MSC treatment significantly reduced mortality rates and the results of this survey supported the decision to proceed to clinical trials that test the effectiveness of MSCs in treating infections causing sepsis in humans [60].

In a phase I clinical trial that included patients admitted to the intensive care unit (ICU) with septic shock, infusion of freshly cultured allogeneic BM-MSCs in doses up to 3 million cells/kg into these ICU patients was shown to be safe as this dose of stem cells did not exacerbate the elevated cytokine levels in the plasma of patients with septic shock [52, 61].

5.2 MSCs in ALI and ARDS

Bacterial pneumonia and sepsis from non-pulmonary causes are the most common etiologies of ALI and ARDS that are associated with mortality rates ranging between 25 and 50% [62–65]. Management of ARDS is mainly supportive with: protective ventilation, fluid conservation, and antimicrobial therapy [62, 64]. In patients with bacterial pneumonia and sepsis, MSCs can attenuate inflammatory process and enhance bacterial clearance [63, 65]. MSCs secrete paracrine factors that can regulate lung permeability and decrease inflammation and this makes MSCs a potentially attractive therapeutic modality for ALI [62]. In patients with ARDS, MSCs can exert beneficial effects by secreting paracrine factors, microvesicles, and transfer of mitochondria. These secretory products have: (1) anti-inflammatory properties that participate in resolving injuries to lung endothelium and alveolar epithelium; (2) regulatory effects on alveolar fluid clearance, thus reducing lung edema; (3) antimicrobial effects mediated by release of antimicrobial factors; and (4) upregulation of monocyte/macrophage phagocytosis [66]. In *Escherichia coli*-injured human lungs, MSCs were able to: restore alveolar fluid clearance, reduce inflammation, and exert antimicrobial activity partly through secretion of keratinocyte growth factor [62].

In patients with bacterial pneumonia causing ALI and ARDS, MSCs could become a promising novel therapeutic modality and an ideal candidate for future cellular therapy due to the following reasons: (1) MSCs are able to differentiate into various cell types, (2) MSCs can secrete multiple bioactive molecules that are capable of stimulating recovery of injured cells and inhibiting inflammation, (3) MSCs lack immunogenicity, and (4) MSCs can perform immunomodulatory functions [62, 63, 65, 67]. In a phase I clinical trial, Jennifer Wilson et al. showed safety of allogeneic BM-MSCs administered to patients with ARDS [56, 68]. However, the role of MSCs in ARDS patients should be carefully evaluated by well-designed multicenter randomized clinical trials [68].

5.3 MSCs in severe and chronic infections

Chronic implant and wound infections that are characterized by biofilm formation are often difficult to treat and they usually require continuous antibiotic therapy for weeks to months. However, alternative therapies for chronically infected wounds include: use of antibiotic impregnated implant materials or biological scaffolds, administration of biofilm disrupting agents, and combining cellular immunotherapy with antibiotics [44].

In patients with very severe aplastic anemia (VSAA), prolonged neutropenia results in refractory and overwhelming bacterial infections as well as invasive fungal infections that are associated with significant morbidity and mortality in these severely immunocompromised individuals [69]. In patients with VSAA lacking human leukocyte antigen identical sibling donors and having refractory infections, co-transplantation of haploidentical HSCs and allogeneic BM-MSCs has been shown to be a safe and a promising therapeutic modality [69].

Studies have shown that: (1) secretion of cathelicidin LL-37 by MSCs could enhance bacterial products indicating that MSCs can upregulate antimicrobial activity in the presence of infection and (2) activated MSCs, when administered intravenously and in combination with conventional antibiotics, can potentially suppress and eradicate chronic *Staphylococcus aureus* biofilm infection in difficult-to-treat locations. Thus, treatment with activated MSCs represents a novel therapeutic option for patients having highly drug-resistant infections [44].

5.4 MSCs in bone, joint, and dental infections

The multidirectional differentiation potential of BM-MSCs is essential for tissue repair after local injury of bones, joints, and medullary adipose tissue. Additionally, the regulation of multiple differentiation potentials of MSCs by various antimicrobial agents affects the recovery from bone and joint infectious diseases [70]. Minocycline induces the following favorable changes in MSCs: migratory capacity, proliferation, gene expression, and growth factor release, ultimately resulting in enhancement of angiogenesis. Also, the triple antimicrobial-loaded hydrogels reduce bacterial bioburden and preserve viability of MSCs in the presence of bacteria [71].

Gingival MSCs encapsulated in silver lactate-containing alginate hydrogel have successfully differentiated into osteogenic tissue and have shown promise for bone tissue engineering with antimicrobial properties against peri-implantitis caused by gram negative bacterial infections [72]. Synthesized antibiotic-containing scaffolds have been shown to possess significantly lower effects on proliferation and viability of human dental pulp stem cells when compared to the saturated ciprofloxacin/metronidazole solution [73].

6. MSCs in viral infections

Studies have shown that: (1) MSCs are susceptible to infection by members of the herpes group of viruses such as: cytomegalovirus, Epstein-Barr virus, herpes simplex virus (HSV) type 1, HSV-2, and VZV, and MSCs become functionally defective following infection with herpes viruses; (2) AT-MSCs can differentiate into functional hepatocyte-like cells but AT-MSCs undergoing hepatic differentiation are not susceptible to infection by hepatitis B virus in vitro; (3) human MSCs are permissive to the highly pathogenic avian influenza A/H5N1 infection and infection of MSCs can cause apoptosis and loss of their immunomodulatory activity; and (4) MSCs can significantly reduce the impairment of alveolar fluid clearance induced by influenza A/H5N1 infection in vitro and prevent or reduce influenza A/H5N1-associated ALI in vivo [28, 34, 74]. The extracellular vesicles (ECVs) secreted by MSCs have anti-inflammatory and anti-influenza properties. Hence, they can be used as cell-free therapy for influenza in humans [75]. Infection of MSCs by respiratory syncytial virus (RSV) alters their immunoregulatory functions by upregulating interferon (IFN)-β and IDO, thus accounting for the lack of protective RSV immunity and for the chronicity of RSV-associated lung diseases such as bronchial asthma and chronic obstructive airway disease [76]. In mice models, treatment with MSCs alleviates inflammation and mortality associated with Japanese encephalitis virus, which is a leading cause of viral encephalitis in Asia [77]. Zika virus infection of human MSCs promotes differential expression of proteins that are linked to several neurological disorders such as Alzheimer's disease, Parkinson's disease, autism, and amyotrophic lateral sclerosis [78].

MSCs exhibit immunomodulatory, anti-inflammatory, and pro-angiogenic properties, and therefore have the potential to improve the outcome of allogeneic HSCT in patients with AA. In a multicenter study that included 75 patients with AA, the combination of HSCs obtained from BM and PB sources as well as MSCs has resulted in amelioration of acute graft versus host disease (GVHD) and viremia resulting ultimately in an improved survival benefit [79].

6.1 MSCs in HIV infection and AIDS

Acquired immunodeficiency syndrome (AIDS), which is caused by human immunodeficiency virus (HIV), poses a real threat to human life [80]. Despite the advent of highly active antiretroviral therapy (HAART) that suppresses plasma viral load but does not cure disease, HIV-1 persists in latent tissue reservoirs, mainly in macrophages and T-helper lymphocytes, and this poses significant challenge to long-term cure [2, 80–82]. HIV-1 predominantly infects HSCs such as macrophages, monocytes, and T-helper lymphocytes [82]. Non-immune responders (NIRs) do respond to HAART, which effectively suppresses HIV replication, but do not show any improvement in their immune status as reflected by an increase in CD4+ T-cell counts [83]. More than 20% of HAART-treated HIV-infected individuals exhibit NIR phenotype and these individuals are at risk of opportunistic infections, cancer, and reduced life expectancy [83].

Coexposure to MSC-conditioned media can enhance the latency-reactivation efficacy of the approved latency reversing drugs vorinostat and panobinostat [81]. Undifferentiated AT resident MSCs are not permissive to HIV-1 infection despite that HIV-1 exposure may increase the expression of some hematopoietic lineage related genes [82]. It has been reported that transfusions of UCB-MSC or more specifically WJ are well tolerated and can efficiently improve immune reconstitution in HIV-infected individuals who are NIRs [83, 84]. Memory CD4 T cells are the key cells organizing all immune actions against HIV while being the targets of HIV infection [85]. MSCs can express receptors that permit their infection by HIV-1. Additionally, human T-lymphotropic virus (HTLV)-1 could infect and replicate in human BM-MSCs possibly by involvement or infiltration of CD4+ lymphocytes [2, 86, 87].

7. MSCs in parasitic infections

Recently, MSCs have been introduced to treat parasitic infections associated with tissue damage in the form of granuloma formation or organ fibrosis such as: schistosomiasis, malaria, and Chagas disease [88, 89]. Studies have shown that MSCs can: (1) ameliorate liver injury and hepatic fibrosis induced by Schistosoma japonicum, particularly when combined with conventional therapies such as praziquantel and (2) play an important role in improving host protective immune responses against malaria by modulating regulatory T cells [88, 89].

7.1 MSCs in Chagas disease

Chagas disease, which is caused by the protozoan *Trypanosoma cruzi*, is endemic in Central and Latin America. However, incidence of the disease has recently increased in the United States of America, Canada, Japan, Australia, and Europe due to migratory movements [2, 90–93]. The disease has acute and chronic phases [90–92]. The acute phase is characterized by intense parasitemia with no or few symptoms while the chronic phase, which extends over indeterminate period of time that may span over years or decades, is characterized by the evolution of cardiac as well as gastrointestinal manifestations reflecting disease complications [90, 91]. Pathogenesis of chronic Chagas cardiomyopathy (CMP) is still debatable but the following have been proposed to be the main pathological mechanisms involved: parasite persistence, microcirculatory alterations, autoimmune mechanisms, and autonomic dysfunction [90, 94]. The cardiac complications of Chagas disease include: myocarditis, dilated CMP, heart failure, arrhythmias, heart block, thromboembolism, stroke, and sudden death [2, 90, 91, 94].

The available and future therapies of Chagas disease include: treatment of arrhythmias and heart failure, antiparasitic therapy, resynchronization treatment, heart transplantation, and stem cell therapies [2, 90, 91, 93, 95]. In patients with chronic Chagas CMP and cardiac failure, conventional pharmacologic therapies are limited by being not always effective, thus rendering the disease incurable [90, 91, 96].

Heart transplantation may occasionally be needed but the procedure has a number of problems including: shortage of donors, high costs, and complications of long-term immunosuppressive therapies administered to recipients of heart transplants [90, 95].

Different stem cell types and delivery approaches have been used in both pre-clinical models as well as clinical trials with the aim of improving cardiac function and reversing complications [95]. In animal models, stem cell therapies have shown reductions in: right ventricular dilatation, and inflammatory infiltrates as well as fibrosis [91, 93]. Stem cell therapy with BM-MSCs has emerged as a novel therapeutic option for Chagas CMP and heart failure [91, 93]. In a murine model of Chagas disease, cotransplantation of autologous BM-MSCs and skeletal myoblasts has been shown to be effective in reversing ventricular dysfunction [94]. Also, in an animal model of chronic Chagas disease, genetic modification of MSCs mobilized by granulocyte colony stimulating factor has increased the immunomodulatory actions and paracrine functions of MSCs by recruitment of suppressor cells such as regulatory T-cells and myeloid-derived suppressor cells [97].

Transplantation of MSCs has shown clinical efficacy in animal or mouse models but studies in humans have not shown equivalent success due to a number of challenges that need to be overcome [2, 91, 93, 95, 98]. In animal models of chronic Chagas CMP, cardiac MSCs have been shown to exert protective effects by decreasing the degrees of fibrosis and inflammatory infiltrates in the affected myocardium [99]. The beneficial effects of MSC therapy in Chagas mice models may be an indirect action of the cells on the heart rather than a direct action of the large numbers of transplanted MSCs on the myocardium [91, 96]. Tracking of infused BM-MSCs in animal models has shown migration of these cells to the heart and their participation in tissue repair or regeneration [91–93]. Although an early clinical trial of intracoronary injection of autologous BM-cells in patients with chronic Chagas CMP and heart failure showed safety and feasibility, a large multicenter, randomized double-blind, placebo-controlled trial using intracoronary infusion of BM-mononuclear cells showed no improvement in cardiac function or in quality of life in patients with chronic Chagas CMP [2, 99, 100].

8. MSCs in tuberculosis

Mycobacterium tuberculosis (MTB) remains a leading cause of morbidity and mortality due to infectious diseases in humans [101]. Multidrug-resistant (MDR) and extensively drug-resistant (XDR) TB, mainly caused by non-adherence to antimicrobial therapy, are recognized health problems in: Eastern Europe, South Africa, and South East Asia [101–103]. Therapeutic strategies that are employed in the management of MDR/XDR TB include: directly observed treatment (DOTS), DOTS-Plus, recombinant human IL-2 by aerosol therapy, and recombinant IFN-γ [102].

Despite the strong host immune response in humans, MTB organisms are capable of persisting or staying dormant for prolonged periods of time, thus resulting in latent infection [104–106]. Hypoxia or hypoxemic microenvironment may favor dormancy of MTB and subsequent evolution of drug resistance [106]. MSCs play a crucial role in the ability of MTB to evade the potent host immune responses and cause TB. Hence, targeting MSCs or nitrous oxide (NO) seems a plausible therapeutic intervention for the design of new effective preventive strategies against TB [107]. Studies have shown that MSCs are recruited into the tuberculous granulomas and they position themselves between the harbored pathogen and the effector T-cells [107–109]. CD271+ BM-MSCs can provide an antimicrobial protective intracellular niche in the host in which dormant MTB can reside for prolonged periods of time [106, 109–111]. MTB infects and persists in a dormant form inside BM-MSCs even after successful antimicrobial therapy [112]. Virulent mycobacteria can manipulate Toll-like receptors and certain signaling pathways including nuclear factor kappa-light-chain-enhancer of activated B cells in order to survive inside the BM stem cells [112]. MSCs can increase NO production in *Mycobacterium abscessus*-infected macrophages through activation of tumor necrosis factor (TNF)-α in the presence of IFN-γ [113]. The cellular crosstalk between TNF-α and prostaglandin-E2 is essential for the increased production of NO in macrophages [113]. Consequently, MSCs may become an ideal choice as adjunct therapy in MDR and XDR TB particularly in individuals with comorbid medical conditions [102, 103, 114]. There are three main clinical trials on the use of MSCs in the treatment of MDR/XDR TB [115–117]. In the first trial, 27 patients with MDR/XDR TB who had been unsuccessfully treated with conventional anti-TB chemotherapy received autologous MSCs, the following results were obtained: all patients showed positive responses to MSC therapy, bacterial discharge from lungs was abolished in 20 patients, tissue damage and lung cavitation resolved in 11 patients, and persistent remission of TB was encountered in 56% of patients after 2 years of autologous MSC transplantation [115]. In the second study, a phase I clinical trial, 36 patients with MDR/XDR TB received anti-TB chemotherapy for 4 weeks; then, they were subjected to autologous MSC transplantation [116]. Six months after autologous transplantation of MSCs: no major adverse events were reported, 70% of patients showed radiological improvement, while 16.7% of patients showed stable radiological appearances. Eighteen months after autologous transplantation of MSCs: 53% of patients were cured, while 10% of patients showed evidence of treatment failure [116]. In the third study, a randomized clinical trial, 72 patients with MDR/XDR TB were included: 36 patients (control group) received conventional anti-TB chemotherapy only, and the other 36 patients (study group) received anti-TB chemotherapy and autologous MSC transplantation [117]. Successful outcomes were encountered in 81% of the study group and 40% of the control group. So, the addition of autologous MSC transplantation to conventional anti-TB chemotherapy significantly enhanced the response rates in patients with MDR/XDR TB [117]. Therefore, combining standard anti-TB chemotherapy with autologous MSC transplantation may ultimately become valuable in increasing the efficacy of anti-TB treatment in patients with MDR-TB [2, 102, 115, 116].

9. MSCs in fungal infections

Administration of human MSCs does not have negative impact on host response against Aspergillus fumigatus [118, 119]. Also, Aspergillus fumigatus does not stimulate MSCs to secrete cytokines that play a major role in the pathogenesis of GVHD indicating that Aspergillus fumigatus is not involved in the pathogenesis of GVHD following HSCT. In an animal model, infusion of BM-MSCs into mice infected with Paracoccidioides brasiliensis failed to induce any antimicrobial effects.

10. Conclusions and future directions

Since their first description in the 1960s, the history of MSCs has witnessed steady progress that ultimately resulted in their clinical application in the treatment of many disorders including several infectious diseases. Although the success has not been uniform with regard to various infections and despite the gap between the achievements in animal studies and results of clinical trials in humans, plenty of efforts have been made to resolve the remaining challenges in the clinical applications of MSCs in several diseases.

Some of the remaining challenges facing the utilization of MSCs in the clinical arena include: (1) encountering failure of treatment or resistance to therapy; (2) the need to have quality control and safety measures; (3) implementation of guidelines and design of specific protocols for: preparation and manufacture, banking and cryopreservation of MSC products, administration and therapeutic use of each type and source of MSCs, and finally tracing of infused MSCs; and (4) performing large prospective multicenter clinical trials on the use of specific MSCs in certain diseases in order to test their uniform efficacy and verify their long-term safety.

Author details

Khalid Ahmed Al-Anazi[1]*, Waleed K. Al-Anazi[2] and Asma M. Al-Jasser[3]

1 Department of Hematology and Hematopoietic Stem Cell Transplantation, Oncology Center, King Fahad Specialist Hospital, Dammam, Saudi Arabia

2 Section of Cytogenetics, Department of Pathology, King Fahad Specialist Hospital, Dammam, Saudi Arabia

3 Department of Research and Studies, General Directorate of Health Affairs in Riyadh Region, Ministry of Health, Riyadh, Saudi Arabia

*Address all correspondence to: kaa_alanazi@yahoo.com

References

[1] Auletta JJ, Deans RJ, Bartholomew AM. Emerging roles for multipotent, bone marrow-derived stromal cells in host defense. Blood. 2012;**119**(8):1801-1809. DOI: 10.1182/blood-2011-10-384354 [Epub: 6 January 2012]

[2] Al-Anazi KA, Al-Jasser AM. Mesenchymal stem cells-their antimicrobial effects and their promising future role as novel therapies of infectious complications in high risk patients. In: Demirer T, editor. Progress in Stem Cell Transplantation. Rijeka: IntechOpen; 2015. DOI: 10.5772/60640

[3] Abdal Dayem A, Lee SB, Kim K, Lim KM, Jeon TI, Seok J, et al. Production of mesenchymal stem cells through stem cell reprogramming. International Journal of Molecular Sciences. 2019;**20**(8):E1922. DOI: 10.3390/ijms20081922

[4] Al-Anazi KA, Bakhit K, Al-Sagheir A, AlHashmi H, Abdulbaqi M, Al-Shibani Z, et al. Cure of insulin-dependent diabetes mellitus by an autologous hematopoietic stem cell transplantation performed to control multiple myeloma in a patient with chronic renal failure on regular hemodialysis. Journal of Stem Cell Biology and Transplantation. 2017;**1**(2):11. DOI: 10.21767/2575-7725.100011

[5] Bobis S, Jarocha D, Majka M. Mesenchymal stem cells: Characteristics and clinical applications. Folia Histochemica et Cytobiologica. 2006; **44**(4):215-230

[6] Kim N, Cho SG. Clinical applications of mesenchymal stem cells. Korean Journal of Internal Medicine. 2013; **28**(4):387-402. DOI: 10.3904/kjim.2013.28.4.387 [Epub: 1 July 2013]

[7] Liu TM. Stemness of mesenchymal stem cells. Preliminary study. Journal of Stem Cell Therapy and Transplantation. 2017;**1**:071-073. DOI: 10.29328/journal.jsctt.1001008

[8] Squillaro T, Peluso G, Galderisi U. Clinical trials with mesenchymal stem cells: An update. Cell Transplantation. 2016;**25**(5):829-848. DOI: 10.3727/096368915X689622 [Epub: 29 September 2015]

[9] Dominici M, Le Blanc K, Mueller I, Slaper-Cortenbach I, Marini F, Krause D, et al. Minimal criteria for defining multipotent mesenchymal stromal cells. The International Society for Cellular Therapy position statement. Cytotherapy. 2006;**8**(4):315-317. DOI: 10.1080/14653240600855905

[10] Nauta AJ, Kruisselbrink AB, Lurvink E, Willemze R, Fibbe WE. Mesenchymal stem cells inhibit generation and function of both CD34+-derived and monocyte-derived dendritic cells. Journal of Immunology. 2006;**177**(4):2080-2087. DOI: 10.4049/jimmunol.177.4.2080

[11] Murray IR, Péault B. Q&A: Mesenchymal stem cells—Where do they come from and is it important? BMC Biology. 2015;**13**:99. DOI: 10.1186/s12915-015-0212-7

[12] Wexler SA, Donaldson C, Denning-Kendall P, Rice C, Bradley B, Hows JM. Adult bone marrow is a rich source of human mesenchymal 'stem' cells but umbilical cord and mobilized adult blood are not. British Journal of Haematology. 2003;**121**(2):368-374. DOI: 10.1046/j.1365-2141.2003.04284.x

[13] Lv FJ, Tuan RS, Cheung KM, Leung VY. Concise review: The surface markers and identity of human mesenchymal stem cells. Stem Cells. 2014;**32**(6):1408-1419. DOI: 10.1002/stem.1681

[14] Kundrotas G. Surface markers distinguishing mesenchymal stem cells

from fibroblasts. Acta Medica Lituanica. 2012;**19**(2):75-79. DOI: 10.6001/actamedica.v19i2.2313

[15] Lin CS, Ning H, Lin G, Lue TF. Is CD34 truly a negative marker for mesenchymal stromal cells? Cytotherapy. 2012;**14**(10):1159-1163. DOI: 10.3109/14653249.2012.729817

[16] Sidney LE, Branch MJ, Dunphy SE, Dua HS, Hopkinson A. Concise review: Evidence for CD34 as a common marker for diverse progenitors. Stem Cells. 2014;**32**(6):1380-1389. DOI: 10.1002/stem.1661

[17] Stzepourginski I, Nigro G, Jacob JM, Dulauroy S, Sansonetti PJ, Eberl G, et al. CD34+ mesenchymal cells are a major component of the intestinal stem cells niche at homeostasis and after injury. Proceedings of the National Academy of Sciences of the United States of America. 2017;**114**(4):E506-E513. DOI: 10.1073/pnas.1620059114 [Epub: 10 January 2017]

[18] Eto H, Ishimine H, Kinoshita K, Watanabe-Susaki K, Kato H, Doi K, et al. Characterization of human adipose tissue-resident hematopoietic cell populations reveals a novel macrophage subpopulation with CD34 expression and mesenchymal multipotency. Stem Cells and Development. 2013;**22**(6):985-997. DOI: 10.1089/scd.2012.0442 [Epub: 21 December 2012]

[19] Alvarez P, Carrillo E, Vélez C, Hita-Contreras F, Martínez-Amat A, Rodríguez-Serrano F, et al. Regulatory systems in bone marrow for hematopoietic stem/progenitor cells mobilization and homing. BioMed Research International. 2013;**2013**:312656. DOI: 10.1155/2013/312656 [Epub: 17 June 2013]

[20] Rochefort GY, Delorme B, Lopez A, Hérault O, Bonnet P, Charbord P, et al. Multipotential mesenchymal stem cells are mobilized into peripheral blood by hypoxia. Stem Cells. 2006;**24**(10):2202-2208 [Epub: 15 June 2006]

[21] Lund TC, Tolar J, Orchard PJ. Granulocyte colony-stimulating factor mobilized CFU-F can be found in the peripheral blood but have limited expansion potential. Haematologica. 2008;**93**(6):908-912. DOI: 10.3324/haematol.12384 [Epub: 9 April 2008]

[22] Gilevich IV, Fedorenko TV, Pashkova IA, Porkhanov VA, Chekhonin VP. Effects of growth factors on mobilization of mesenchymal stem cells. Bulletin of Experimental Biology and Medicine. 2017;**162**(5):684-686. DOI: 10.1007/s10517-017-3687-0 [Epub: 31 March 2017]

[23] Xu L, Li G. Circulating mesenchymal stem cells and their clinical implications. Journal of Orthopaedic Translation. 2014;**2**(1):1-7. DOI: 10.1016/j.jot.2013.11.002

[24] Koning JJ, Kooij G, de Vries HE, Nolte MA, Mebius RE. Mesenchymal stem cells are mobilized from the bone marrow during inflammation. Frontiers in Immunology. 2013;**4**:49. DOI: 10.3389/fimmu.2013.00049 [eCollection 2013]

[25] Ding SSL, Subbiah SK, Khan MSA, Farhana A, Mok PL. Empowering mesenchymal stem cells for ocular degenerative disorders. International Journal of Molecular Sciences. 2019;**20**(7):E1784. DOI: 10.3390/ijms20071784

[26] Mansoor H, Ong HS, Riau AK, Stanzel TP, Mehta JS, Yam GH. Current trends and future perspective of mesenchymal stem cells and exosomes in corneal diseases. International Journal of Molecular Sciences. 2019;**20**(12):E2853. DOI: 10.3390/ijms20122853

[27] Leyendecker A Jr, Pinheiro CCG, Amano MT, Bueno DF. The use of human mesenchymal stem cells as

therapeutic agents for the in vivo treatment of immune-related diseases: A systematic review. Frontiers in Immunology. 2018;**9**:2056. DOI: 10.3389/fimmu.2018.02056 [eCollection 2018]

[28] Thanunchai M, Hongeng S, Thitithanyanont A. Mesenchymal stromal cells and viral infection. Stem Cells International. 2015;**2015**:860950. DOI: 10.1155/2015/860950 [Epub: 29 July 2015]

[29] Yang K, Wang J, Wu M, Li M, Wang Y, Huang X. Mesenchymal stem cells detect and defend against gammaherpesvirus infection via the cGAS-STING pathway. Scientific Reports. 2015;**5**:7820. DOI: 10.1038/srep07820

[30] Azadniv M, Myers JR, McMurray HR, Guo N, Rock P, Coppage ML, et al. Bone marrow mesenchymal stromal cells from acute myelogenous leukemia patients demonstrate adipogenic differentiation propensity with implications for leukemia cell support. Leukemia. 2019. DOI: 10.1038/s41375-019-0568-8 [Epub ahead of print]

[31] Pinho S, Lacombe J, Hanoun M, Mizoguchi T, Bruns I, et al. PDGFRα and CD51 mark human nestin+ sphere-forming mesenchymal stem cells capable of hematopoietic progenitor cell expansion. The Journal of Experimental Medicine. 2013;**210**(7):1351-1367. DOI: 10.1084/jem.20122252 [Epub: 17 June 2013]

[32] Pleyer L, Valent P, Greil R. Mesenchymal stem and progenitor cells in normal and dysplastic hematopoiesis-masters of survival and clonality? International Journal of Molecular Sciences. 2016;**17**(7):E1009. DOI: 10.3390/ijms17071009

[33] Shi C. Recent progress toward understanding the physiological function of bone marrow mesenchymal stem cells. Immunology. 2012;**136**(2): 133-138. DOI: 10.1111/j.1365-2567.2012.03567.x

[34] Al-Anazi KA, Al-Anazi WK, Al-Jasser AM. The beneficial effects of varicella zoster virus. Journal of Hematology and Clinical Research. 2019;**3**:016-049. DOI: 10.29328/journal.jhcr.1001010

[35] Kyurkchiev D, Bochev I, Ivanova-Todorova E, Mourdjeva M, Oreshkova T, Belemezova K, et al. Secretion of immunoregulatory cytokines by mesenchymal stem cells. World Journal of Stem Cells. 2014;**6**(5): 552-570. DOI: 10.4252/wjsc.v6.i5.552

[36] Krampera M, Cosmi L, Angeli R, Pasini A, Liotta F, Andreini A, et al. Role for interferon-gamma in the immunomodulatory activity of human bone marrow mesenchymal stem cells. Stem Cells. 2006;**24**(2):386-398. DOI: 10.1634/stemcells.2005-0008 [Epub: 25 August 2005]

[37] Castro-Manrreza ME, Montesinos JJ. Immunoregulation by mesenchymal stem cells: Biological aspects and clinical applications. Journal of Immunological Research. 2015;**2015**:394917. DOI: 10.1155/2015/394917 [Epub: 19 April 2015]

[38] Avanzi S, Leoni V, Rotola A, Alviano F, Solimando L, Lanzoni G, et al. Susceptibility of human placenta derived mesenchymal stromal/stem cells to human herpesviruses infection. PLoS One. 2013;**8**(8):e71412. DOI: 10.1371/journal.pone.0071412. Print 2013

[39] Abdi Z, Eskandary H, Nematollahi-Mahani SN. Effects of two types of human cells on outgrowth of human glioma in rats. Turkish Neurosurgery. 2018;**28**(1):19-28. DOI: 10.5137/1019-5149.JTN.18697-16.1

[40] Dong HJ, Li G, Meng HP, Shang CZ, Luo Y, Wen G, et al. How can

mesenchymal stem cells penetrate the blood brain barrier? Turkish Neurosurgery. 2018;**28**(6):1013-1014. DOI: 10.5137/1019-5149.JTN.22639-18.1

[41] Conaty P, Sherman LS, Naaldijk Y, Ulrich H, Stolzing A, Rameshwar P. Methods of mesenchymal stem cell homing to the blood-brain barrier. Methods in Molecular Biology. 1842; **2018**:81-91. DOI: 10.1007/978-1-4939-8697-2_6

[42] Liu L, Eckert MA, Riazifar H, Kang DK, Agalliu D, Zhao W. From blood to the brain: can systemically transplanted mesenchymal stem cells cross the blood-brain barrier? Stem Cells International. 2013;**2013**:435093. DOI: 10.1155/2013/435093 [Epub: 12 August 2013]

[43] Christodoulou I, Goulielmaki M, Devetzi M, Panagiotidis M, Koliakos G, Zoumpourlis V. Mesenchymal stem cells in preclinical cancer cytotherapy: A systematic review. Stem Cell Research and Therapy. 2018;**9**(1):336. DOI: 10.1186/s13287-018-1078-8

[44] Johnson V, Webb T, Norman A, Coy J, Kurihara J, Regan D, et al. Activated mesenchymal stem cells interact with antibiotics and host innate immune responses to control chronic bacterial infections. Scientific Reports. 2017;7(1):9575. DOI: 10.1038/s41598-017-08311-4

[45] Alcayaga-Miranda F, Cuenca J, Khoury M. Antimicrobial activity of mesenchymal stem cells: Current status and new perspectives of antimicrobial peptide-based therapies. Frontiers in Immunology. 2017;**8**:339. DOI: 10.3389/fimmu.2017.00339 [eCollection 2017]

[46] Sutton MT, Fletcher D, Ghosh SK, Weinberg A, van Heeckeren R, Kaur S, et al. Antimicrobial properties of mesenchymal stem cells: therapeutic potential for cystic fibrosis infection, and treatment. Stem Cells International. 2016;**2016**:5303048. DOI: 10.1155/2016/5303048 [Epub: 26 January 2016]

[47] Valdivia-Silva J, Medina-Tamayo J, Garcia-Zepeda EA. Chemokine-derived peptides: Novel antimicrobial and antineoplasic agents. International Journal of Molecular Sciences. 2015; **16**(6):12958-12985. DOI: 10.3390/ijms160612958

[48] Laroye C, Gibot S, Reppel L, Bensoussan D. Concise review: Mesenchymal stromal/stem cells: A new treatment for sepsis and septic shock? Stem Cells. 2017;**35**(12):2331-2339. DOI: 10.1002/stem.2695 [Epub: 16 September 2017]

[49] Laroye C, Lemarié J, Boufenzer A, Labroca P, Cunat L, Alauzet C, et al. Clinical-grade mesenchymal stem cells derived from umbilical cord improve septic shock in pigs. Intensive Care Medicine Experimental. 2018;**6**(1):24. DOI: 10.1186/s40635-018-0194-1

[50] Johnson CL, Soeder Y, Dahlke MH. Concise review: Mesenchymal stromal cell-based approaches for the treatment of acute respiratory distress and sepsis syndromes. Stem Cells Translational Medicine. 2017;**6**(4):1141-1151. DOI: 10.1002/sctm.16-0415 [Epub: 9 January 2017]

[51] Skirecki T, Mikaszewska-Sokolewicz M, Godlewska M, Dołęgowska B, Czubak J, Hoser G, et al. Mobilization of stem and progenitor cells in septic shock patients. Scientific Reports. 2019;**9**(1):3289. DOI: 10.1038/s41598-019-39772-4

[52] McIntyre LA, Stewart DJ, Mei SHJ, Courtman D, Watpool I, Granton J, et al., Canadian Critical Care Trials Group; Canadian Critical Care Translational Biology Group. Cellular immunotherapy for septic shock. A phase I clinical trial. The American Journal of Respiratory and Critical Care Medicine. 2018;**197**(3):

337-347. DOI: 10.1164/rccm.201705-1006OC

[53] Guillamat-Prats R, Camprubí-Rimblas M, Bringué J, Tantinyà N, Artigas A. Cell therapy for the treatment of sepsis and acute respiratory distress syndrome. Annals of Translational Medicine. 2017;**5**(22):446. DOI: 10.21037/atm.2017.08.28

[54] Li J, Huang S, Wu Y, Gu C, Gao D, Feng C, et al. Paracrine factors from mesenchymal stem cells: A proposed therapeutic tool for acute lung injury and acute respiratory distress syndrome. International Wound Journal. 2014; **11**(2):114-121. DOI: 10.1111/iwj.12202 [Epub: 26 December 2013]

[55] Saeedi P, Halabian R, Fooladi AAI. Antimicrobial effects of mesenchymal stem cells primed by modified LPS on bacterial clearance in sepsis. Journal of Cellular Physiology. 2019;**234**(4): 4970-4986. DOI: 10.1002/jcp.27298 [Epub: 14 September 2018]

[56] Wilson JG, Liu KD, Zhuo H, Caballero L, McMillan M, Fang X, et al. Mesenchymal stem (stromal) cells for treatment of ARDS: A phase 1 clinical trial. The Lancet Respiratory Medicine. 2015;**3**(1):24-32. DOI: 10.1016/S2213-2600(14)70291-7 [Epub: 17 December 2014]

[57] Laroye C, Boufenzer A, Jolly L, Cunat L, Alauzet C, Merlin JL, et al. Bone marrow vs Wharton's jelly mesenchymal stem cells in experimental sepsis: A comparative study. Stem Cell Research & Therapy. 2019;**10**(1):192. DOI: 10.1186/s13287-019-1295-9

[58] Kim H, Darwish I, Monroy MF, Prockop DJ, Liles WC, Kain KC. Mesenchymal stromal (stem) cells suppress pro-inflammatory cytokine production but fail to improve survival in experimental staphylococcal toxic shock syndrome. BMC Immunology. 2014;**15**:1. DOI: 10.1186/1471-2172-15-1

[59] Asano K, Yoshimura S, Nakane A. Adipose tissue-derived mesenchymal stem cells attenuate Staphylococcal enterotoxin A-induced toxic shock. Infection and Immunity. 2015;**83**(9): 3490-3496. DOI: 10.1128/IAI.00730-15 [Epub: 22 June 2015]

[60] Lalu MM, Sullivan KJ, Mei SH, Moher D, Straus A, Fergusson DA, et al. Evaluating mesenchymal stem cell therapy for sepsis with preclinical meta-analyses prior to initiating a first-in-human trial. eLife. 2016;**5**: e17850. DOI: 10.7554/eLife.17850

[61] Schlosser K, Wang JP, Dos Santos C, Walley KR, Marshall J, Fergusson DA, et al., Canadian Critical Care Trials Group and the Canadian Critical Care Translational Biology Group. Effects of mesenchymal stem cell treatment on systemic cytokine levels in a phase 1 dose escalation safety trial of septic shock patients. Critical Care Medicine. 2019;**47**(7):918-925. DOI: 10.1097/CCM.0000000000003657

[62] Lee JW, Krasnodembskaya A, McKenna DH, Song Y, Abbott J, Matthay MA. Therapeutic effects of human mesenchymal stem cells in ex vivo human lungs injured with live bacteria. The American Journal of Respiratory and Critical Care Medicine. 2013;**187**(7):751-760. DOI: 10.1164/rccm.201206-0990OC

[63] Sung DK, Chang YS, Sung SI, Yoo HS, Ahn SY, Park WS. Antibacterial effect of mesenchymal stem cells against *Escherichia coli* is mediated by secretion of beta- defensin- 2 via toll- like receptor 4 signalling. Cellular Microbiology. 2016; **18**(3):424-436. DOI: 10.1111/cmi.12522 [Epub: 27 October 2015]

[64] Wang YY, Li XZ, Wang LB. Therapeutic implications of mesenchymal stem cells in acute lung injury/acute respiratory distress syndrome. Stem Cell Research &

Therapy. 2013;**4**(3):45. DOI: 10.1186/scrt193

[65] Morrison T, McAuley DF, Krasnodembskaya A. Mesenchymal stromal cells for treatment of the acute respiratory distress syndrome: The beginning of the story. Journal of the Intensive Care Society. 2015;**16**(4): 320-329. DOI: 10.1177/1751143715586420 [Epub: 21 May 2015]

[66] Laffey JG, Matthay MA. Fifty years of research in ARDS. Cell-based therapy for acute respiratory distress syndrome. Biology and potential therapeutic value. The American Journal of Respiratory and Critical Care Medicine. 2017; **196**(3):266-273. DOI: 10.1164/rccm.201701-0107CP

[67] Hayes M, Curley G, Laffey JG. Mesenchymal stem cells - a promising therapy for acute respiratory distress syndrome. F1000 Medicine Reports. 2012;**4**:2. DOI: 10.3410/M4-2 [Epub: 3 January 2012]

[68] Zhang GY, Liao T, Zhou SB, Fu XB, Li QF. Mesenchymal stem (stromal) cells for treatment of acute respiratory distress syndrome. The Lancet Respiratory Medicine. 2015;**3**(4):e11-e12. DOI: 10.1016/S2213-2600(15)00049-1

[69] Yue C, Ding Y, Gao Y, Li L, Pang Y, Liu Z, et al. Cotransplantation of haploidentical hematopoietic stem cells and allogeneic bone marrow-derived mesenchymal stromal cells as a first-line treatment in very severe aplastic anemia patients with refractory infections. Europian Journal of Haematology. 2018; **100**(6):624-629. DOI: 10.1111/ejh.13060 [Epub: 25 April 2018]

[70] Li H, Yue B. Effects of various antimicrobial agents on multi-directional differentiation potential of bone marrow-derived mesenchymal stem cells. World Journal of Stem Cells. 2019;**11**(6):322-336. DOI: 10.4252/wjsc.v11.i6.322

[71] Guerra AD, Rose WE, Hematti P, Kao WJ. Minocycline enhances the mesenchymal stromal/stem cell pro-healing phenotype in triple antimicrobial-loaded hydrogels. Acta Biomaterialia. 2017;**51**:184-196. DOI: 10.1016/j.actbio.2017.01.021 [Epub: 7 January 2017]

[72] Diniz IM, Chen C, Ansari S, Zadeh HH, Moshaverinia M, Chee D, et al. Gingival mesenchymal stem cell (GMSC) delivery system based on RGD-coupled alginate hydrogel with antimicrobial properties: A novel treatment modality for peri-implantitis. Journal of Prosthodontics. 2016;**25**(2): 105-115. DOI: 10.1111/jopr.12316 [Epub: 27 July 2015]

[73] Kamocki K, Nör JE, Bottino MC. Dental pulp stem cell responses to novel antibiotic-containing scaffolds for regenerative endodontics. International Endodontic Journal. 2015;**48**(12): 1147-1156. DOI: 10.1111/iej.12414 [Epub: 24 December 2014]

[74] Chan MC, Kuok DI, Leung CY, Hui KP, Valkenburg SA, Lau EH, et al. Human mesenchymal stromal cells reduce influenza A H5N1-associated acute lung injury in vitro and in vivo. Proceedings of the National Academy of Sciences of the United States of America. 2016;**113**(13):3621-3626. DOI: 10.1073/pnas.1601911113 [Epub: 14 March, 2016]

[75] Khatri M, Richardson LA, Meulia T. Mesenchymal stem cell-derived extracellular vesicles attenuate influenza virus-induced acute lung injury in a pig model. Stem Cell Research & Therapy. 2018;**9**(1):17. DOI: 10.1186/s13287-018-0774-8

[76] Cheung MB, Sampayo-Escobar V, Green R, Moore ML, Mohapatra S, Mohapatra SS. Respiratory syncytial virus-infected mesenchymal stem cells regulate immunity via interferon beta

and indoleamine-2,3-dioxygenase. PLoS One. 2016;**11**(10):e0163709. DOI: 10.1371/journal.pone.0163709 [eCollection 2016]

[77] Bian P, Ye C, Zheng X, Yang J, Ye W, Wang Y, et al. Mesenchymal stem cells alleviate Japanese encephalitis virus-induced neuroinflammation and mortality. Stem Cell Research & Therapy. 2017;**8**(1):38. DOI: 10.1186/s13287-017-0486-5

[78] Beys-da-Silva WO, Rosa RL, Santi L, Berger M, Park SK, Campos AR, et al. Zika virus infection of human mesenchymal stem cells promotes differential expression of proteins linked to several neurological diseases. Molecular Neurobiology. 2019;**56**(7):4708-4717. DOI: 10.1007/s12035-018-1417-x [Epub: 30 October 2018]

[79] Chen M, Zheng Z, Hu J, Yang T. Co-transplantation of mesenchymal stem cells can ameliorates acute GVHD and viremia after allo-HSCT for aplastic anemia: A multi-center retrospective study of 75 patients. HemaSphere. 2019; **3**:710. Poster Session II: Stem cell transplantation—Clinical. PS1537. DOI: 10.1097/01.HS9.0000564408.36600.07

[80] Halder UC. Bone marrow stem cells to destroy circulating HIV: A hypothetical therapeutic strategy. Journal of Biological Research-Thessaloniki. 2018;**25**:3. DOI: 10.1186/s40709-018-0075-5 [eCollection: 29 December 2018]

[81] Chandra PK, Gerlach SL, Wu C, Khurana N, Swientoniewski LT, Abdel-Mageed AB, et al. Mesenchymal stem cells are attracted to latent HIV-1-infected cells and enable virus reactivation via a non-canonical PI3K-NFκB signaling pathway. Scientific Reports. 2018;**8**:14702. DOI: 10.1038/s41598-018-32657-y

[82] Nazari-Shafti TZ, Freisinger E, Roy U, Bulot CT, Senst C, Dupin CL, et al. Mesenchymal stem cell derived hematopoietic cells are permissive to HIV-1 infection. Retrovirology. 2011; **8**(1):3. DOI: 10.1186/1742-4690-8-3

[83] Allam O, Samarani S, Ahmad A. Mesenchymal stem cell therapy in HIV-infected HAART-treated nonimmune responders restores immune competence. AIDS. 2013;**27**(8):1349-1352. DOI: 10.1097/QAD.0b013e32836010f7

[84] Zhang Z, Fu J, Xu X, Wang S, Xu R, Zhao M, et al. Safety and immunological responses to human mesenchymal stem cell therapy in difficult-to-treat HIV-1-infected patients. AIDS. 2013;**27**(8):1283-1293. DOI: 10.1097/QAD.0b013e32835fab77

[85] Zhang J, Crumpacker C. Eradication of HIV and cure of AIDS, now and how? Frontiers in Immunology. 2013;**4**:337. DOI: 10.3389/fimmu.2013.00337

[86] Cotter EJ, Chew N, Powderly WG, Doran PP. HIV type 1 alters mesenchymal stem cell differentiation potential and cell phenotype ex vivo. AIDS Research and Human Retroviruses. 2011;**27**(2):187-199. DOI: 10.1089/aid.2010.0114 [Epub: 7 October 2010]

[87] Rodrigues ES, de Macedo MD, Pinto MT, Orellana MD, Rocha Junior MC, de Magalhães DA, et al. HTLV-1 infects human mesenchymal stromal cell in vitro and modifies their phenotypic characteristics. Virology. 2014;**449**:190-199. DOI: 10.1016/j.virol.2013.11.022 [Epub: 6 December 2013]

[88] Zhang Y, Mi JY, Rui YJ, Xu YL, Wang W. Stem cell therapy for the treatment of parasitic infections: Is it far away? Parasitology Research. 2014; **113**(2):607-612. DOI: 10.1007/s00436-013-3689-4 [Epub: 26 November 2013]

[89] Xu H, Qian H, Zhu W, Zhang X, Yan Y, Mao F, et al. Mesenchymal stem

cells relieve fibrosis of Schistosoma japonicum-induced mouse liver injury. Experimental Biology and Medicine (Maywood, NJ). 2012;(5):237, 585-592. DOI: 10.1258/ebm.2012.011362

[90] de Carvalho AC, Carvalho AB. Stem cell-based therapies in Chagasic cardiomyopathy. Biomed Research International. 2015;**2015**:436314. DOI: 10.1155/2015/436314 [Epub: 15 June 2015]

[91] Jasmin, Jelicks LA, Koba W, Tanowitz HB, Mendez-Otero R, et al. Mesenchymal bone marrow cell therapy in a mouse model of chagas disease. Where do the cells go? PLoS Neglected Tropical Diseases. 2012; **6**(12):e1971. DOI: 10.1371/journal.pntd.0001971

[92] Souza BS, Azevedo CM, d Lima RS, Kaneto CM, Vasconcelos JF, Guimarães ET, et al. Bone marrow cells migrate to the heart and skeletal muscle and participate in tissue repair after *Trypanosoma cruzi* infection in mice. International Journal of Experimental Pathology. 2014;**95**(5):321-329. DOI: 10.1111/iep.12089 [Epub: 30 June 2014]

[93] Irion CI, Paredes BD, Brasil GV, Cunha STD, Paula LF, Carvalho AR, et al. Bone marrow cell migration to the heart in a chimeric mouse model of acute Chagasic disease. Memórias do Instituto Oswaldo Cruz. 2017;**112**(8): 551-560. DOI: 10.1590/0074-02760160526

[94] Guarita-Souza LC, Carvalho KA, Woitowicz V, Rebelatto C, Senegaglia A, Hansen P, et al. Simultaneous autologous transplantation of cocultured mesenchymal stem cells and skeletal myoblasts improves ventricular function in a murine model of Chagas disease. Circulation. 2006;**114**(1 Suppl): I120-I1124. DOI: 10.1161/CIRCULATIONAHA.105.000646

[95] de Carvalho KA, Abdelwahid E, Ferreira RJ, Irioda AC, Guarita-Souza LC. Preclinical stem cell therapy in Chagas disease: Perspectives for future research. World Journal of Transplantation. 2013;**3**(4):119-126. DOI: 10.5500/wjt.v3.i4.119

[96] Jasmin, Jelicks LA, Tanowitz HB, Peters VM, Mendez-Otero R, de Carvalho ACC, et al. Molecular imaging, biodistribution and efficacy of mesenchymal bone marrow cell therapy in a mouse model of Chagas disease. Microbes and Infection. 2014;**16**(11): 923-935. DOI: 10.1016/j.micinf.2014.08.016 [Epub: 16 September 2014]

[97] Silva DN, Souza BSF, Vasconcelos JF, Azevedo CM, Valim CXR, Paredes BD, et al. Granulocyte-colony stimulating factor-overexpressing mesenchymal stem cells exhibit enhanced immunomodulatory actions through the recruitment of suppressor cells in experimental Chagas disease cardiomyopathy. Frontiers in Immunology. 2018;**9**:1449. DOI: 10.3389/fimmu.2018.01449 [eCollection 2018]

[98] Ribeiro Dos Santos R, Rassi S, Feitosa G, Grecco OT, Rassi A Jr, da Cunha AB, et al., Chagas Arm of the MiHeart Study Investigators. Cell therapy in Chagas cardiomyopathy (Chagas arm of the multicenter randomized trial of cell therapy in cardiopathies study): A multicenter randomized trial. Circulation. 2012; **125**(20):2454-2461. DOI: 10.1161/CIRCULATIONAHA.111.067785 [Epub: 20 April 2012]

[99] Silva DN, de Freitas Souza BS, Azevedo CM, Vasconcelos JF, Carvalho RH, Soares MB, et al. Intramyocardial transplantation of cardiac mesenchymal stem cells reduces myocarditis in a model of chronic Chagas disease cardiomyopathy. Stem Cell Research & Therapy. 2014;**5**(4):81. DOI: 10.1186/scrt470

[100] Vilas-Boas F, Feitosa GS, Soares MB, Mota A, Pinho-Filho JA, Almeida AJ, et al. Early results of bone marrow cell transplantation to the myocardium of patients with heart failure due to Chagas disease. Arquivos Brasileiros de Cardiologia. 2006;**87**(2): 159-166. DOI: 10.1590/s0066-782x2006001500014

[101] Joshi L, Chelluri LK, Gaddam S. Mesenchymal stromal cell therapy in MDR/XDR tuberculosis: A concise review. Archivum Immunologiae et Therapiae Experimentalis. 2015;**63**(6): 427-433. DOI: 10.1007/s00005-015-0347-9

[102] Iyer RN, Chelluri EP, Chelluri LK. Role of mesenchymal stem cell based therapies in MDR/XDR TB and co-morbidities. Journal of Stem Cell Research & Therapy. 2015;**5**:284. DOI: 10.4172/2157-7633.1000284

[103] Khan FN, Zaidi KU, Thawani V. Stem cell therapy: An adjunct in the treatment of mdr tuberculosis. Journal of Stem Cell Research & Therapeutics. 2017;**3**(3):259-261. DOI: 10.15406/jsrt.2017.03.00099

[104] Sikri K, Tyagi JS. The evolution of Mycobacterium tuberculosis dormancy models. Current Science. 2013;**105**(5): 607-616

[105] Tornack J, Reece ST, Bauer WM, Vogelzang A, Bandermann S, Zedler U, et al. Human and mouse hematopoietic stem cells are a depot for dormant Mycobacterium tuberculosis. PLoS One. 2017;**12**(1):e0169119. DOI: 10.1371/journal.pone.0169119 [eCollection 2017]

[106] Garhyan J, Bhuyan S, Pulu I, Kalita D, Das B, Bhatnagar R. Preclinical and clinical evidence of Mycobacterium tuberculosis persistence in the hypoxic niche of bone marrow mesenchymal stem cells after therapy. The American Journal of Pathology. 2015;**185**(7): 1924-1934. DOI: 10.1016/j.ajpath.2015.03.028 [Epub: 8 June 2015]

[107] Mittal R. Mesenchymal stem cells: the new players in the pathogenesis of tuberculosis. Journal of Microbial & Biochemical Technology. 2011;**03**(03). DOI: 10.4172/1948-5948.100000e3

[108] Raghuvanshi S, Sharma P, Singh S, Van Kaer L, Das G. Mycobacterium tuberculosis evades host immunity by recruiting mesenchymal stem cells. Proceedings of the National Academy of Sciences of the United States of America. 2010;**107**(50):21653-21658. DOI: 10.1073/pnas.1007967107 [Epub: 6 December 2010]

[109] Al-Anazi KA, Al-Jasser AM, Alsaleh K. Infections caused by Mycobacterium tuberculosis in recipients of hematopoietic stem cell transplantation. Frontiers in Oncology. 2014;**4**:231. DOI: 10.3389/fonc.2014.00231 [eCollection 2014]

[110] Das B, Kashino SS, Pulu I, Kalita D, Swami V, Yeger H, et al. CD271(+) bone marrow mesenchymal stem cells may provide a niche for dormant *Mycobacterium tuberculosis*. Science Translational Medicine. 2013;**5**(170): 170ra13. DOI: 10.1126/scitranslmed.3004912

[111] Beamer G, Major S, Das B, Campos-Neto A. Bone marrow mesenchymal stem cells provide an antibiotic-protective niche for persistent viable *Mycobacterium tuberculosis* that survive antibiotic treatment. The American Journal of Pathology. 2014;**184**(12):3170-3175. DOI: 10.1016/j.ajpath.2014.08.024 [Epub: 16 October 2014]

[112] Naik SK, Padhi A, Ganguli G, Sengupta S, Pati S, Das D, et al. Mouse bone marrow Sca-1+ CD44+ mesenchymal stem cells kill avirulent Mycobacteria but not *Mycobacterium tuberculosis* through modulation of

cathelicidin expression via the p38 mitogen-activated protein kinase-dependent pathway. Infection and Immunity. 2017;**85**(10). DOI: 10.1128/IAI.00471-17. Print October 2017. pii: e00471-17

[113] Kim JS, Cha SH, Kim WS, Han SJ, Cha SB, Kim HM, et al. A novel therapeutic approach using mesenchymal stem cells to protect against *Mycobacterium abscessus*. Stem Cells. 2016;**34**(7):1957-1970. DOI: 10.1002/stem.2353 [Epub: 27 March 2016]

[114] Arora VK, Dhot PS, Singhal P. Stem cells in MDR-TB and XDR-TB. Current Respiratory Medicine Reviews. 2014;**10**(4):238-240. DOI: 10.2174/1573398X11666150109223613

[115] Erokhin VV, Vasil'eva IA, Konopliannikov AG, Chukanov VI, Tsyb AF, Bagdasarian TR, et al. Systemic transplantation of autologous mesenchymal stem cells of the bone marrow in the treatment of patients with multidrug resistant pulmonary tuberculosis. Problemy Tuberkuleza i Boleznei Legkikh. 2008;**10**:3-6

[116] Skrahin A, Ahmed RK, Ferrara G, Rane L, Poiret T, Isaikina Y, et al. Autologous mesenchymal stromal cell infusion as adjunct treatment in patients with multidrug and extensively drug-resistant tuberculosis: An open-label phase 1 safety trial. The Lancet Respiratory Medicine. 2014;**2**(2):108-122. DOI: 10.1016/S2213-2600(13)70234-0 [Epub: 9 January 2014]

[117] Skrahin AE, Jenkins HE, Hurevich H, Solodovnokova V, Isaikina Y, Klimuk D, et al. Potential role of autologous mesenchymal stromal cells in the treatment of multidrug and extensively drug-resistant tuberculosis. European Respiratory Journal. 2016;**48**:PA1919. DOI: 10.1183/13993003.congress-2016.PA1919

[118] Schmidt S, Tramsen L, Schneider A, Schubert R, Balan A, Degistirici Ö, et al. Impact of human mesenchymal stromal cells on antifungal host response against *Aspergillus fumigatus*. Oncotarget. 2017;**8**(56):95495-95503. DOI: 10.18632/oncotarget.20753. [eCollection: 10 November 2017]

[119] Arango JC, Puerta-Arias JD, Pino-Tamayo PA, Salazar-Peláez LM, Rojas M, González Á. Impaired anti-fibrotic effect of bone marrow-derived mesenchymal stem cell in a mouse model of pulmonary paracoccidioidomycosis. PLoS Neglected Tropical Diseases. 2017;**11**(10):e0006006. DOI: 10.1371/journal.pntd.0006006. [eCollection: October 2017]

Chapter 4

The Role of Mesenchymal Stromal Cells in the Management of Osteoarthritis of the Knee

Charan Thej and Pawan Kumar Gupta

Abstract

Osteoarthritis (OA) is one of the most common chronic, inflammatory, and degenerative diseases affecting the synovial joints, the hip, and the knee. OA is commonly managed clinically by treating pain with anti-inflammatory medicines using nonsteroidal anti-inflammatory drugs (NSAIDS) or analgesics. In severe OA patients, invasive knee replacement surgery is the last option. Treatment of OA using mesenchymal stromal cells (MSCs) has been widely explored due to their anti-inflammatory properties and chondrogenic differentiation potential. In this chapter, we comprehensively discuss in detail the in vitro OA potency development, OA preclinical studies, and clinical trials conducted using MSCs.

Keywords: osteoarthritis, pooled human bone marrow-derived mesenchymal stromal cells, potency assay, preclinical studies, clinical studies

1. Introduction

Common factors linked to osteoarthritis (OA) occurrence are increasing age (>55 years) and obesity [1]. The gender also seems to play a major role, where the majority of OA patients are women and higher prevalence has been liked to menopause. Radiological evidence suggests that about 70% of women above the age of 65 years are affected by OA [2, 3]. Other factors such as genetic predisposition, extrinsic environmental factors, nutrition, and lack of exercise are reasons for the increased prevalence of OA. It has been reported by the World Health Organization (WHO) that 10–15% of the populations aged >60 years exhibit a certain degree of OA [4]. It has been reported by the National Health Portal of India that 22–39% of the Indian population are affected by OA. As reported by the United Nations Organization (UNO), 130 million people will be affected by OA with over 40 million people with severe disability due to disease progression [3].

The etiology of OA is believed to be multifactorial. Some of the main reasons include the biomechanical disease progression due to the narrowing of space in the joints, bone hypertrophy, and formation of new osteophytes in the articular margins causing stiffness and pain in the joints. In addition, an imbalance in the synthesis and release of cytokines by chondrocytes in the disease state could be the main reason for the continual inflammatory state in the joint. During the initial stages of OA, catabolic interleukins (IL) such as IL-1α and IL-1β and tumor necrosis factor α (TNFα) increase inflammation affecting cartilage metabolism and homeostasis. TNFα is a proinflammatory cytokine implicated in the degradation of matrix proteins synthesized by chondrocytes and synoviocytes [5]. Further, increase in the levels of interferon γ (IFNγ) in the joint worsens the inflammatory state and structure of the joint leading to degradation of proteoglycans such as sulfated glycosaminoglycans (sGAG) [5, 6].

2. Current treatment options for osteoarthritis

Currently, pain in OA is pharmacologically managed using nonsteroidal anti-inflammatory drugs (NSAIDS), opioids, and analgesics. Corticosteroid injections have also been used for relieving severe pain in OA patients. Recent attempts have been made to use TNFα blockers as recent studies have proven the significant role of TNFα in contribution to the pathogenesis of OA [7]. Research by several groups has implicated the role of nerve growth factor (NGF) and its binding to tropomyosin receptor kinase A (trk A) which leads to downstream signaling and activation of peripheral and central pain molecules causing severe pain. The therapeutic efficiency of anti-NGF antibodies to block NGF or its antagonists has been studied by several groups for relieving pain. The pain-relieving effects of anti-NGF antibodies fasinumab and fulranumab manufactured by Regeneron Pharmaceuticals and Janssen Pharmaceutica, respectively, have been evaluated in phase III clinical trials [8]. In addition to pain relief, efforts have been made to halt further cartilage damage using slow-acting symptomatic drugs such as chondroitin sulfate and glucosamine sulfate. Orally administered chondroitin and glucosamine have shown to relieve joint pain equivalently compared to NSAIDs. These molecules, intact or broken, could be absorbed into the matrix of the joint and prevent cartilage degeneration. Although glucosamine and chondroitin sulfate have been clinically proven to be safe, their therapeutic efficacy in protecting the cartilage matrix was found to the variable [9]. In grade 4 OA (Kellgren and Lawrence classifications), patients are advised to opt for total knee replacement surgery [10]. Alternatively, autologous chondrocyte implantation (ACI) has been suggested and reported to be successful. In the ACI method, the chondrocytes from patients are taken, culture-expanded in vitro, and then implanted back into the knees of patients. This procedure is invasive and has a lesser success rate than total knee replacement surgeries [11].

Apart from ACI, the efficacy of autologous platelet-rich plasma (PRP) in providing pain relief and promoting cartilage regeneration has been recently investigated by several groups [12]. The PRP is rich in platelets that secrete several growth factors and cytokines such as platelet-derived growth factor (PDGF), hepatocyte growth factor (HGF), insulin-like growth factor 1 (IGF-1), vascular endothelial growth factor (VEGF), and prostaglandin E2 (PGE-2) [13]. Several research groups have reported that intra-articular injections of PRP primarily reduced inflammation mediated by PGE-2, HGF, and IGF-1. IGF-1 synthesized and secreted by platelets is shown to prevent leukocyte infiltration into the joint space, thereby reducing the levels of IL-1β and TNFα in the synovial fluid [13]. Overall intra-articular injection of PRP has been shown to maintain joint homeostasis. However, clinical trial data suggest that the effect of PRP seems to last for only 3 weeks and thereafter reduces. The symptoms of OA were seen to relapse after a period of 1 year. Although promising results were observed using PRP in the hydrogel, chitosan, or hyaluronic acid (HA) scaffolds [14], efficacy is yet to be shown in elaborate randomized clinical trials (RCTs).

3. Mesenchymal stromal cells

The history of mesenchymal stromal/stem cells (MSCs) dates back to 1960 when seminal studies conducted by Friedenstein showed the isolation of MSCs from bone marrow (BM) which were capable of forming ectopic bone in vivo. This was found to be a non-hematopoietic fibroblast-like, colony-forming cell which primarily supported hematopoietic stem cells in the perivascular niche [15]. Owen and Friedenstein discovered that these cells were capable of differentiating into the osteogenic lineage [16]. Subsequently, the multipotent plasticity of that bone marrow MSCs (BMMSCs) was identified and shown that they were capable of differentiating into osteocytes, chondrocytes, and adipocytes in vitro [17]. In addition to the abovementioned three lineages, Caplan and colleagues demonstrated that these cells were capable of differentiating into cells of the muscle, tendons/ligaments, and connective tissue after which he coined the term "mesenchymal stem cells" [18]. Bianco and Gehron Robey deduced that *cbfa1* gene was the master regulator for directing the osteogenic fate of MSCs. Because of the ability of MSCs to form osteocytes, they named them skeletal stem cells [19]. In 2006, the International Society for Cellular Therapy (ISCT) proposed the name multipotent mesenchymal stromal cells and defined that MSCs must adhere to the criteria of being plastic adherent; express surface markers CD105, CD73, and CD90; lack the expression of hematopoietic markers CD45, CD34, CD14 or CD11b, CD79α or CD19, and HLA-DR; and differentiate into osteoblasts, chondrocytes, and adipocytes under suitable conditions in vitro [20]. In addition to their differentiation capacity, MSCs have been shown to elicit immunosuppressive and immunomodulatory effects on T lymphocytes, B cells, dendritic cells (DC), and natural killer (NK) cells either by cell-cell interactions or by secretion of anti-inflammatory molecules such as indoleamine 2,3-dioxygenase (IDO) and prostaglandin E2 (PGE-2), interleukin-4 (IL-4), interleukin-10 (IL-10), and transforming growth factor β (TGFβ) making them ideal cell types for treatment of diseases [21–23]. Because of their ability to differentiate into chondrocytes in vitro and with their anti-inflammatory and immunomodulatory functions, they were believed to be candidate cell type to treat diseases such as OA. MSCs have been isolated from over 18 different tissue sources. The most commonly used tissue sources for isolating MSCs apart from bone marrow are the adipose tissue, umbilical cord, placenta, and dental pulp. However, autologous or allogeneic BMMSCs are currently the most widely used cell type in clinical trials for various disease indications. They are considered the "gold standard" MSC type because of their extensive characterization that took place for over 5 decades.

4. Possible mechanism of action (MoA) of BMMSCs for treatment of osteoarthritis

The pathophysiology of OA is characterized by degradation of hyaline cartilage causing narrowing of joint space leading to subchondral sclerosis, subchondral cysts, hypertrophic chondrocytes, and formation of osteophytes. The friction caused by the rubbing of joints results in chronic pain in OA patients [24]. Degeneration of cartilage extracellular matrix (ECM) may be caused due to the increase in the levels of proteolytic enzymes such as matrix metalloproteases (MMPs) and aggrecanases mediated by IL-1β and TNFα [25]. BMMSCs express a wide range of properties that are anticipated to be beneficial for treating genetic, mechanical, and age-related degeneration in diseases such as OA. In our previous publication, we have in detail attempted to deduce the possible mechanism of action (MoA) of allogeneic pooled BMMSC population [25]. Briefly, BMMSCs are known to be immunomodulatory in nature, primarily because of their potential to significantly suppress the proliferation of inflammatory T cells, monocytes, and dendritic cells either by direct cell-to-cell contact. In addition, they secrete a wide range of anti-inflammatory molecules such as PGE-2, IDO, IL1Ra, and IL-10 [26, 27]. BMMSCs influence the local osteoarthritic microenvironment by stimulating resident chondrogenic progenitor cells and promote their differentiation into mature chondrocytes mediated by secretion of bone morphogenetic proteins (BMPs) and TGFβ1 [28]. BMMSCs are known to differentiate into chondrocytes in vitro using differentiation cues such as BMP-7 and TGFβ1. A similar mechanism could be involved in the differentiation of BMMSCs in vivo. With the increase in the levels of BMP-7 and TGFβ1 in the local joint milieu, mediated by a change in expression of master regulatory genes such as Sox9, HoxA, HoxD, and Gli3, BMMSCs could differentiate into chondrogenic progenitor cells (CPCs) in vivo. The CPCs further differentiate into chondroblasts characterized by definitive upregulation of collagen types II B, IX, and XI. Subsequently, the CPCs differentiate into mature chondrocytes regulated by balanced expression of collagen X (Col X) and synthesize the secretion of collagen II which is made of sGAG building blocks which maintain the structural integrity of hyaline cartilage [25]. Very high expression of collagen X has been linked to hypertrophy of chondrocytes and formation of fibrous cartilage, and thus a regulated expression of Col X would likely result in deposition of hyaline cartilage [29]. From the above-described multimodal MoA, it is clear that BMMSCs are an ideal cell population which could contribute significantly for an effective treatment of OA.

5. Advantages of using a pooled human BMMSC (phBMMSC, Stempeucel®) product for treating osteoarthritis

In the current therapeutic scenario, the common practice is to screen several individual donors, isolate MSCs, and characterize them based on their key characteristics such as their surface marker expression, tri-lineage differentiation potential, and immunomodulatory and paracrine properties [30–32]. It is inevitable that a product that is manufactured using a master cell bank (MCB) made from a single donor will result in exhaustion. Successively, a product that is made using another single donor MSC bank, although presumably similar in basic characteristics qualifying the identity and safety criteria, may not have the same functional attributes which may lead to varied therapeutic outcomes. Eminent scientific groups have demonstrated donor-to-donor variability in properties of MSCs such as their clonogenicity, growth kinetics, and differentiation potential [33]. A comparative analysis of five different BMMSC populations showed significant variation in the proteomic profile of these cells. Only 13% similarity in the proteomic profile which included transcriptional and translational regulators, kinases, receptor proteins, and cytokines between the five BMMSC populations was found. A maximum of 72% similarity in the proteome was observed between two of the five analyzed cell populations [34]. Disparities in clinical trial outcomes have been reported where BMMSCs derived from single donors have been used. A steroid-refractory acute graft-versus-host disease (SR-aGvHD) clinical trial conducted in both children (n = 25) and adults (n = 30) using BMMSC products derived from 92 HLA-matched and HLA-mismatched donors resulted in only 50% overall durable complete response, while the remaining patients did not respond or partially responded to the treatment [35]. Similar variations with limited response rates were observed in a phase III GvHD trial conducted by Osiris Therapeutics using Prochymal® with only 35% complete response rate compared to 30% in the placebo arm [36]. It has been suggested that improper selection of a BM donor and making a single donor-derived cell product could lead to substantial variations in therapeutic outcomes [37]. In order to challenge this issue, some scientific groups have suggested pooling of BMMSCs from two or more donors in order to compensate for the variation and balance the properties between different donor cell populations. Samuelsson et al. showed that a two- or three-donor pooled BMMSC product could

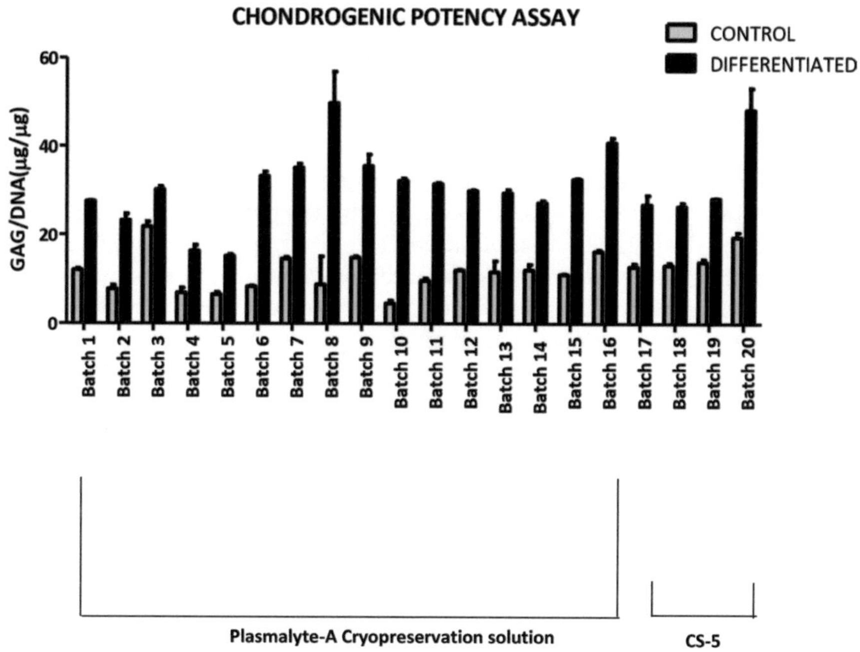

Figure 1.
Chondrogenic potency assessment using quantification of sGAG in 16 batches of Stempeucel® cryopreserved in PlasmaLyte A-based cryopreservation solution and 4 batches of Stempeucel® cryopreserved in CS 5.

optimize the immunosuppressive properties of these cells in vitro [38]. Later, Kuçi et al. showed substantial variability in the immunosuppressive properties of individual donor-derived BMMSCs (n = 8). On the contrary, a mesenchymal end product (MEP) made by pooling BMMNCs from eight donors resulted in a cell population that consistently suppressed an MLR in vitro [39]. Subsequently, they went on to conduct a multicentric SR-aGvHD clinical trial in 51 children and 18 adults using MEP/MSC Frankfurt am Main (MSC-FFM, Obnitix®) cells and observed 83% overall response (complete response, 32%; partial response, 51%) [40]. At Stempeutics Research Pvt. Ltd., we were the first group to develop an allogeneic pooled human BMMSC product called Stempeucel® using an established, robust pooling protocol and a two-tier manufacturing and banking system as previously described [41, 42]. Recently, we have published our comprehensive studies including in vitro chondrogenic properties and preclinical and clinical findings establishing the efficacy and safety of using Stempeucel® for the treatment of OA of the knee joint [43]. In this study, we found that several manufactured batches of Stempeucel®, when differentiated into the chondrocyte lineage, downregulated the expression of the gene Sox9 and upregulated the expression of collagen type 2A (Col2A) gene confirming their differentiation into the chondrogenic lineage. The same Stempeucel® batches synthesized substantial levels of sGAG (30 ± 1.8 µg/µg GAG/DNA) which were estimated using a dimethylmethylene blue-based biochemical assay kit (**Figure 1**). These properties indicate that Stempeucel® could be a potential treatment option for treating OA.

6. Development of a potency assay for Stempeucel® intended to treat osteoarthritis

The US Food and Drug Administration (USFDA) describes potency assays as "The specific ability or capacity of the product, as indicated by appropriate laboratory tests or by adequately controlled clinical data obtained through the administration of the product in the manner intended, to effect a given result" (US-FDA, 21 CFR 600). For any cell therapy product (CTP) intended to be used for a particular indication, a specific, quantifiable, potency test or array must be developed. The development of a potency assay must begin with in vitro and preclinical studies based on the MoA of the CTP. The confirmation of the assay or the identified marker must be evaluated in every large-scale manufactured batch of the CTP during the progress of the phase I and phase II clinical studies. A quantifiable range for the potency test must be defined and implemented during the course of phase III clinical trial [44]. In order to predict the efficacy of a CTP, either in vitro biochemical assays or biological assays or in vivo biological assessment could be implemented. For example, a company called TiGenix (Leuven, Belgium) has developed and adopted an assay matrix where an ex vivo polymerase chain reaction (PCR) array for autologous chondrocytes (ChondroCelect) is performed and ectopic cartilage formation is correlated to the histology sections of an orthotopic goat model where ChondroCelect is implanted [45, 46]. Jeong et al. have demonstrated that thrombospondin-2 (TSP-2) could be an effective marker to predict the chondrogenic efficiency of umbilical cord-derived MSCs (UC-MSCs). They demonstrated that UC-MSCs, through the TSP-2 secretion, can promote chondrogenesis via PKCa, ERK, P38/MAPK, and Notch signaling pathways [47]. Recently, another group estimated the levels of TSP-2 to evaluate the chondrogenic potency of a UC-MSC product (Cellistem®OA, Cells for Cells, Brazil) intended to be used in phase I/phase II RCT for knee OA [48]. Other scientific groups have shown that autologous culture-expanded chondrocytes could be embedded in collagen-1 and injected subcutaneously in nude mice to predict the potency of several bioactive molecules in promoting chondrogenesis [49]. For the first time, we have developed a chondrogenic potency assay for an allogeneic pooled bone marrow-derived MSC product (phBMMSCs, Stempeucel®). Preliminarily, we culture-expanded and differentiated several Stempeucel® batches into the chondrogenic lineage using commercially available differentiation assay kits (Thermo Fisher Scientific, USA). To confirm the differentiation, we evaluated the Col2A mRNA expression in differentiated cells and compared them with the undifferentiated control cells. After observing a significant increase in the Col2A expression of differentiated cells, we enzymatically digested both the differentiated and undifferentiated cells to quantify the levels of sGAG synthesized by these cells using a 1,9-dimethylmethylene blue (DMMB)-based assay kit (Blyscan, Biocolor, UK). We further normalized the levels of sGAG with the amount of DNA from the same number of cells. We evaluated the sGAG levels in 20 batches of Stempeucel® of which 16 batches were cryopreserved in our older formulation (10% dimethyl sulfoxide (DMSO), 5% human serum albumin (HSA) and PlasmaLyte A) and also four batches of Stempeucel® cryopreserved in a new cGMP grade CryoStor 5 solution (CS5, BioLife Solutions). We observed a significant and consistent increase in the levels of sGAG in the differentiated cells compared to the undifferentiated cells (undifferentiated, 11.9 ± 4.6 GAG/DNA (μg/μg); differentiated, 31 ± 8.6 GAG/DNA (μg/μg; $P < 0.0001$; $n = 20$)) (**Figure 1**). Based on our results, we propose that the sGAG assay is a simple, quantifiable, and robust potency assay which could also be a part of a bigger potency assay matrix to predict the chondrogenic potency of therapeutic cells intended to treat cartilage defects.

7. Preclinical efficacy studies in OA

Many studies have demonstrated that MSCs are nontoxic and non-tumorigenic when tested in various animal models [50, 51]. Prior to evaluating the efficacy of Stempeucel® in an appropriate preclinical model of OA, we had earlier evaluated the preclinical safety and toxicity of Stempeucel® in rodent and non-rodent

Author/ year	Animal	OA model	Cell type and dose	Vehicle	Study duration time points	Reference
Murphy et al. (2003)	Goat	ACLT-meniscectomy	10×10^6 Autologous (goat) BMMSC + HA	HA	12 and 26 weeks	[65]
Frisbie et al. (2009)	Horse	Arthroscopic surgery	10.5×10^6 Autologous (horse) BMMSC	Saline	10 weeks	[61]
Sato et al. (2012)	Pig	Spontaneous	7×10^6 Xenogeneic (human) BMMSC	HA/PBS	1, 3, and 5 weeks	[66]
Song et al. (2014)	Sheep	ACLT-meniscectomy	10×10^6 Autologous (sheep) BMMSC	PBS	8 weeks	[63]
Delling et al. (2015)	Sheep	Bilateral meniscectomy	20×10^6 Autologous (sheep) BMMSC	PBS	0, 1, 4, 8, and 12 weeks	[64]
Singh et al. (2014)	Rabbit	ACLT	1×10^6 Autologous (rabbit) BMMSC	Culture medium	4 and 6 weeks	[55]
Chiang et al. (2016)	Rabbit	ACLT	1×10^6 Allogeneic (rabbit) BMMSC	HA	6 and 12 weeks	[56]
Diekman et al. (2013)	Mouse	Closed tibial plateau fracture	1×10^5 Allogeneic (mice) BMMSC	Saline/mouse albumin	8 weeks	[52]
Suhaeb et al. (2012)	Rat	MIA injection	3.5×10^6 Allogeneic (rat) BMMSC	HA	3 and 9 weeks	[67]
Kim et al. (2014)	Rat	ACLT-meniscectomy	1×10^6 Allogeneic (rat) BMMSC	Culture medium	3 and 6 weeks	[53]
Yang et al. (2015)	Rat	ACLT-meniscectomy	0.5×10^6 Autologous (rat) BMMSC	PBS	3 weeks	[54]
Gupta et al. (2016)	Rat	MIA injection	0.6×10^6 or 1.3×10^6 Xenogeneic (human) pooled BMMSC	PlasmaLyte A	4, 8, and 12 weeks	[43]

Table 1.
OA preclinical studies using BMMSCs.

models. In the same study, we evaluated the feasibility of multiple routes of cell injection. Tumorigenic analysis in severe combined immunodeficient (SCID) mice showed that Stempeucel® is non-tumorigenic. In addition, the biodistribution kinetics of CM-DiI labeled Stempeucel® in the systemic circulation and also in muscle tissue were studied in both rats and mice [51].

It is important to demonstrate the efficacy of any cell therapy product in an animal model of disease before administrating the product in humans with the same disease. It is imperative to determine the suitability of using animal stem cells in animals or human stem cells in immunocompromised/immunocompetent animals. A common regulatory requirement is to have animal data for the same test product that is intended to be tested in humans. In our recently published work, we evaluated the efficacy of Stempeucel® in a monosodium iodoacetate (MIA)-induced

OA model in Wistar rats. We demonstrated the dose-dependent efficacy of two Stempeucel® doses of 0.65×10^6 (25×10^6 human equivalent dose, HED) and 1.3×10^6 (50×10^6 HED) followed by an injection of hyaluronic acid (HA). A significant dose-dependent reduction in pain scores was observed in both low and high Stempeucel® doses compared to the HA alone and disease control group. Histological evaluation of joint tissue sections in all study groups showed significant improvement in proteoglycan staining in both low and high Stempeucel® administered groups indicating significant regeneration of the cartilage in both groups compared to the HA alone and disease control groups [43].

Similar to the animal model we used, other scientific groups have created articular cartilage defects in small animals, such as mice [52], rats [43, 53, 54], and rabbits [55, 56]. Smaller animal models are cost-effective and easy to house, and rodents are available in a variety of genetically modified strains with minimal biological variability [57, 58]. However, the small joint size, thin cartilage, altered biomechanics, and increased spontaneous intrinsic healing hamper the study of the regenerative capacity of stem cells and these mechanisms of healing which cannot be fully extrapolated to human cartilage repair [59, 60]. Rodents have mainly been used to assess the chondrogenesis of cell-based therapies by subcutaneous, intramuscular, and intra-articular implantations of cells [60]. Of all small animals, the rabbit model is the most utilized model in cartilage regeneration studies because of the slightly larger knee joint size than rodents [55, 56]. Despite their limited translational capacity, small animals can be very useful as a proof-of-principle study and to assess therapy safety before moving on to preclinical studies using larger animals [60].

Large animal models play a more substantial role in translational research because of a larger joint size and thicker cartilage; however, their preclinical use is often hindered by high costs and difficulties in animal handling. A variety of large animal models have been used to investigate cartilage repair strategies, including horses [61], dogs [62], sheep [63, 64], goats [65], and pigs [66], each with their own strengths and limitations. We have listed some relevant published studies which have used autologous, allogeneic, or xenogeneic BMMSCs to treat OA induced by various methods (**Table 1**).

Based on the positive efficacy outcomes of our preclinical study, subsequently, we demonstrated the safety and optimal dose for efficacy in a phBMMSC product, Stempeucel®, in a randomized, double-blind, placebo-controlled dose-finding phase II clinical trial in Indian patients [43].

8. Clinical trials in osteoarthritis of the knee joint

8.1 Safety of mesenchymal stromal cells in clinical trials

Lalu MM et al. conducted a systematic review of clinical trials that examined the use of MSCs to evaluate their safety [68]. A total of 36 studies having 1012 participants with different clinical conditions was evaluated. Eight studies were randomized control trials (RCTs) and enrolled 321 participants. Only prospective clinical trials that used the intravenous or intra-arterial route of administration in different age groups were analyzed. Meta-analysis did not detect an association with MSC administration and acute infusional toxicity, organ system complications, infection, and death. There was a significant association between MSCs and transient fever at or shortly after MSC administration which was not associated with long-term sequelae. Most importantly, the meta-analysis showed no serious adverse event due to the administration of MSCs and specifically found no association between MSCs and tumor formation. In another study, Peeters et al. [69] did a systemic review of the safety of intra-articular administration of culture-expanded stem cells. A total of 844 procedures (mean follow-up of 21 months) was analyzed. Four SAEs were reported—one infection following bone marrow aspiration (BMA) that resolved with antibiotics, one pulmonary embolism after 2 weeks of BMA, and two adverse events not related to the therapy. Other adverse events documented were increased pain/swelling and dehydration after BMA. In another review, a recent analysis of adverse events (AEs) in 2372 orthopedic patients treated with autologous stem cell therapies and followed up for 2.2 years has been published [70]. The common AEs reported included post-procedure pain and pain due to progressive degenerative joint disease in under 4% of the population. Hence, we can conclude that the systemic administration of MSC including intra-articular administration is safe.

8.2 Efficacy of stem cells including mesenchymal stromal cells in clinical trials of osteoarthritis of the knee joint

Several clinical trials have been conducted using bone marrow mononuclear cells, adipose tissue-derived stromal vascular fraction (AD-SVF), adipose tissue-derived mesenchymal stromal cells (AD-MSCs), or bone marrow-derived mesenchymal stromal cells (BMMSCs) in OA of the knee joint. The list of the published clinical trials in chronological order is given in **Table 2**. Administration of the cells has been fairly standardized, with the cells being administered either directly intra-articularly or under ultrasound guidance. Few trials have been conducted using the arthroscopic method of administration with direct implantation of the cells alone or with a scaffold at the site of cartilage injury.

The first clinical study has been published way back in 2002 by Wakitani et al. [71]. In this study of 12 patients who underwent high tibial osteotomy, BMMSCs at a dose of 13 million cells were embedded in collagen gel and transplanted into the cartilage defect and covered with autologous periosteum. The clinical improvement was not significantly different from the control group, but the arthroscopic and histological evidence was better in the transplanted group than the control arm. Since then many studies have been published, but still many contentious issues regarding cell therapy in OA are being discussed. We will try to discuss a few burning issues in this chapter:

a. **Level of evidence regarding the use of MSC therapy in OA**: Jevotovsky et al. [106] did a systemic review of 61 studies to look at the study evidence level, MSC protocol, treatment results, and AEs. The levels of evidence were defined by Marx et al. stating the level of evidence as level 1, randomized controlled trial; level II, prospective cohort study or observational study with dramatic effects; level III, retrospective cohort study or case–control study; level IV, case series; and level V, mechanism-based reasoning [107]. These levels of evidence help physicians to come to clinical decisions. In this review, a total of 2390 patients in 61 studies was identified. Most of the studies used adipose-derived stem cells (ADSCs) (n = 29) or bone marrow-derived stem cells (BMSCs) (n = 30). The majority of the studies (57%) were level IV evidence which consists of therapeutic case series without comparative groups. Only five and nine studies were level I and level II evidence, respectively, in a total of 288 patients. Additionally, 11% were level III retrospective cohort studies, and 8% were level V single-patient case reports. The published data highlights the need for more level I and level II evidence to evaluate the role of MSC treatment in OA patients. However, the majority of the studies have reported positive results and an association between MSC therapy and symptomatic and radiologic improvement in these patients.

Author/year	Sample size	Study design	Grade of OA	Cell type and dose	Control group	Outcome measures	Outcomes	Follow-up period	Reference
Wakitani (2002)	24	A single-arm control study	Stage I to stage II Ahlback changes	Auto BMMSC, 13×10^6 cells embedded in soluble collagen (n = 12)	High tibial osteotomy (n = 12)	Hospital for Special Surgery knee rating scale, arthroscopy and histology	No significant difference in clinical evaluations between the two groups, arthroscopy and histology showed partially hyaline cartilage-like tissue	16 months	[71]
Centeno et al. (2008)	1	Case study	—	Auto BMMSC, 2.4×10^6 cells + 1 ml NC + 1 ml PRP	Nil	VAS, ROM, and MRI	Decreased VAS pain scores; increased the range of motion; MRI, statistically significant cartilage and meniscus growth	6 months	[72]
Haleem et al. (2010)	5	Case series	Outerbridge grade III or grade IV	Auto BMMSC + PR-FG, 15×10^6 cells	Nil	Lysholm and RHSSK scores and X-rays and MRI	Lysholm and RHSSK scores showed statistically significant improvement, MRI of three patients revealed complete defect fill	12 months	[73]
Nejadnik et al. (2010)	72	Cohort study	Lesion grade 3 or grade 4	Auto BMMSC	n = 12 each in BMMSC and chondrocyte group	ICRS, SF-36, IKDC, Lysholm knee scale and Tegner activity level scale	SF-36 showed physical role functioning improvement in the BMMSC group, no difference in other outcome measures	24 months	[74]
Davatchi et al. (2011)	4	Single-arm study	Moderate to severe OA	Auto BMMSC, $8–9 \times 10^6$ cells	Nil	VAS pain score, time to walk and number of stairs to climb to produce pain, the resting time to induce the gelling pain, ROM and patellae crepitus	Walking time for the pain to appear improved for three patient, VAS pain score and number of stairs to climb improved for all patients, improvement in crepitus	6 months	[75]
Saw et al. (2011)	5	Single-arm study	ICRS grade 3 or grade 4	Auto PBPC (8 ml) + HA/ weekly injection × 5	Nil	Arthroscopy and histology	Arthroscopy showed articular cartilage regeneration and histologically showed hyaline cartilage	26 months	[76]

Author/year	Sample size	Study design	Grade of OA	Cell type and dose	Control group	Outcome measures	Outcomes	Follow-up period	Reference
Koh et al. (2012)	25	Case–control study	KL grade 3 or grade 4	Arthroscopic debridement + Auto AD-MSC, 1.89×10^6 cells + PRP	Arthroscopic debridement + PRP	Lysholm score, Tegner activity scale, and VAS scores	The clinical scores preoperatively were significantly poorer than those of the control group but at the last follow-up visit were similar and not significantly different between the two groups	18 months	[77]
Orozco et al. (2013)	12	Case series	KL grade 2 to grade 4	Auto BMMSC, 40×10^6 cells	Nil	VAS score, Lequesne indexes, WOMAC scores, MRI T2 mapping	All clinical scores decreased significantly, MRI T2 mapping showed improvement of cartilage quality	12 months	[78]
Koh et al. (2013)	18	Case series	KL grade 3 or grade 4	Auto AD-MSC, 1.18×10^6 cells + PRP	Nil	WOMAC score, Lysholm score, Tegner activity scale, and VAS scores, MRI (WORMS score)	WOMAC, Lysholm, Tegner, and VAS scores improved significantly, WORMS score in MRI improved significantly	26 months	[79]
Van Pham et al. (2014)	21	Case series	KL grade 2 or grade 3	Auto SVF	Nil	VAS, Lysholm score, and MRI	VAS scores improved, Lysholm scores increased, MRI showed increased cartilage thickness	8.5 months	[80]
Koh et al. (2014)	37 knees (35 patients)	Case series	KL grade 1 or grade 2	Auto AD-MSC, 3.8×10^6 cells	Nil	IKDC score, Tegner activity scale, and cartilage repair assessed using ICRS grading	IKDC and Tegner activity scale scores significantly improved, ICRS grades showed 2 of the 37 lesions (5%) were grade I (normal) and 7 (19%) were grade II (near normal)	26.5 months	[81]
Koh et al. (2015)	30	Case series	KL grade 2 or grade 3	Auto SVF + PRP, 42×10^6 cells	Nil	Lysholm score, KOOS, VAS score, arthroscopic evaluation (n = 16)	Significant improvement in all clinical outcomes, 87.5% of patients (14/16) improved or maintained cartilage status	24 months	[82]

Author/year	Sample size	Study design	Grade of OA	Cell type and dose	Control group	Outcome measures	Outcomes	Follow-up period	Reference
Munar et al. (2015)	50	Case series	KL grade 2 to grade 4	Auto BMMSC, 40×10^6 cells	Nil	VAS, Lequesne and WOMAC indices, MRI (T2 mapping)	All clinical scores improved, T2 mapping, PCI decreased significantly	12 months	[83]
Davatchi et al. (2016)	4	Case series	Moderate to severe OA	Auto BMMSC, $8–9 \times 10^6$ cells	Nil	VAS pain score, time to walk and number of stairs to climb to produce pain, the resting time to induce the gelling pain, ROM and patellae crepitus	All parameters still better than baseline at 5 years follow-up for three patients	60 months	[84]
Soler et al. (2016)	15	Single-arm, open-label phase I/phase II trial	KL grade 2 or grade 3	Auto BMMSC, 41×10^6 cells	Nil	VAS score, questionnaire, QOL SF-36 questionnaire, Lequesne functional index and WOMAC score, MRI (T2 mapping)	The clinical scores improved, SF-36 showed improvement of parameters, T2 mapping showed signs of cartilage regeneration	12 months	[85]
Sampson et al. (2016)	125	Retrospective case series	KL grade 3 or grade 4	Auto BMC + PRP (8 weeks apart)	Nil	VAS score, patient satisfaction scale	VAS score and patient satisfaction score improved in all patients	4.8 months	[86]
Fodor and Paulseth (2016)	6 patients (8 knees)	Case series	KL grade 1 to grade 3	Auto SVF, 14.1×10^6 cells	Nil	VAS score, WOMAC score, ROM, TUG test, and MRI	VAS and WOMAC scores significantly improved, ROM and TUG improved, MRI showed no detectable structural differences	12 months	[87]
Kim et al. (2016)	20 patients (24 knees)	Case series	KL grade 1 or grade 2	Auto AD-MSC, 4.4×10^6 cells	Nil	IKDC score, Tegner activity scale, MRI MOAKS and MOCART score	Clinical outcomes significantly improved, MOAKS and MOCART score significantly improved	24 months	[88]

Author/ year	Sample size	Study design	Grade of OA	Cell type and dose	Control group	Outcome measures	Outcomes	Follow-up period	Reference
Pak et al. (2016)	3	Case reports	KL grade 3	Auto AD-SVF + HA + PRP, PRP repeated weekly × 3	Nil	FRI, ROM and VAS score, MRI	All clinical scores improved in three patients, MRI showed cartilage-like tissue regeneration	4.5 months	[89]
Pers et al. (2016)	18	Open-label, phase I study	KL grade 3 or grade 4	Auto AD-MSC, low (2×10^6), medium (10×10^6), and high dose (50×10^6), 6 patients each	Nil	Safety, WOMAC, VAS, PGA, SAS and KOOS index, MRI, dGEMRIC in 6 patients	Safety established, low dose most effective, and all parameters improved as compared to baseline, dGEMRIC improved in three patients	6 months	[90]
Koh et al. (2016)	80	RCT	ICRS grade 3 or grade 4	Auto AD-MSC + fibrin glue + microfracture, 5×10^6 cells (group 1)	n = 40 (group 2) (microfracture)	Lysholm score, KOOS, VAS score, MRI, cartilage repair tissue scoring system, arthroscopy and histology	MRI, better signal intensity for repair tissue in group 1 (80%) as compared to 72.5% in group 2; KOOS pain and symptom subscores, significantly greater for group 1; arthroscopy and histology, no significant difference	24 months	[91]
Gupta et al. (2016)	60	Double-blind, phase II, RCT	KL grade 2 or grade 3	Allo BMMSC + HA, four doses (25, 50, 75, 150 × 10^6 cells)	n = 20 (placebo + HA)	Safety, VAS, ICOAP, WOMAC, MRI, WORMS score	Safety established; AE were predominant in the higher-dose groups; VAS, ICOAP, and WOMAC scores best in the lowest dose; MRI, no significant difference	24 months	[43]
Lamo-Espinosa et al. (2016)	30	Phase I/phase II, RCT	KL grade 2 to grade 4	Auto BMMSC + HA, two doses (10 and 100 × 10^6 cells)	n = 10, (HA)	Safety, VAS score, WOMAC, MRI, WORMS	Safety established. VAS, WOMAC, and WORMS scores significant in high-dose group at 12 months follow-up	12 months	[92]
de Windt et al. (2017)	10	Phase I/phase II single-center study	Modified Outerbridge grade 3 or grade 4	Allo BMMSCs +10 or 20% autologous chondrons	Nil	Safety, KOOS, VAS, MRI, second-look arthroscopy, histology	No SUSAR; KOOS and VAS scores improved significantly; MRI, showed complete filling of the defect; arthroscopy, effective defect fill, and integration in the surrounding tissue; histology, positive staining for both type I and type II collagen and proteoglycan	12 months	[93]

Author/ year	Sample size	Study design	Grade of OA	Cell type and dose	Control group	Outcome measures	Outcomes	Follow-up period	Reference
Turajane et al. (2017)	60	RCT, single-center study	KL grade 1 to grade 3	3 groups, 20 each; first group (AAPBSC + PRP + G-CSF+ HA + MCS); second group (AAPBSC + PRP + HA + MCS); third group (control) (all given weekly × 3 injections)	20 patients, IA, HA alone	Avoidance of TKA intervention and WOMAC scores	TKA done in three patients in the control group but not in the cell group; WOMAC, all groups reached statistically significant improvements within the individual (intra) groups	12 months	[94]
Shapiro et al. (2017)	25 patients, 50 knees	RCT, single-blind, placebo-controlled	Bilateral OA, KL grade 1 to grade 3	25 knees; 5 ml of Auto BMAC +10 ml of platelet-poor bone marrow plasma	25 knees; sterile saline, 15 ml	ICOAP, VAS scores, MRI, T2 mapping	No SAE, patients had a similar decrease in scores in VAS and ICOAP scores in both BMAC- and saline-treated arthritic knees	6 months	[95]
Park et al. (2017)	7	Open-label, single-arm, phase I/ phase II	KL grade 3 and ICRS grade 4	Two doses; Allo hUCB MSCs and HA hydrogel	Nil	ICRS cartilage repair, VAS, IKDC, MRI, histological findings	VAS and IKDC improved at 24 weeks and stable till 7 years; histology at 1 year showed hyaline-like cartilage; MRI at 3 years showed the persistence of regenerated cartilage	7 years	[96]
Nguyen et al. (2017)	30	Placebo-controlled trial	KL grade 2 or grade 3	15 patients; AM + Auto AD (SVF + PRP; 10^7 SVF cells/ml)	15 patients, AM	Safety, WOMAC, Lysholm, and modified VAS scores, MRI	Safety established, WOMAC scores not significant between two arms at 6 and 12 months but significant at 18 months; increased Lysholm and VAS scores in the treatment group compared with the placebo; MRI, MRI demonstrated cartilage layer was thicker in the treatment group	18 months	[97]
Pintat et al. (2017)	19	Single-arm study	Patellofemoral OA	IA AD-MSC + PRP	Nil	WOMAC, MRI, T2 mapping	WOMAC scores significantly lower in treatment arm than baseline; MRI, no change	12 months	[98]

Author/year	Sample size	Study design	Grade of OA	Cell type and dose	Control group	Outcome measures	Outcomes	Follow-up period	Reference
Russo et al. (2017)	30	Single-arm study	KL grade 1 to grade 3, grade > II (ICRS classification)	Auto microfragmented adipose tissue	Nil	VAS, KOOS, IKCD, subjective, Tegner Lysholm knee	IKDC and KOOS, improvement of 20 points; VAS and Tegner Lysholm score, improvement in 24 and 31 points, respectively	12 months	[99]
Yokota et al. (2017)	13	Single-arm study	KL grade 3 or grade 4	IA Auto AD- SVF; 2.5 ml SVF containing 3×10^7 SVF cells/knee	Nil	VAS, JKOM, WOMAC	Scores improved by an average of 35% over baseline for JKOM, 32% improvement in WOMAC, and 40% for pain (VAS)	6 months	[100]
Jo et al. (2017)	18	Single-arm study	Knee OA	Auto AD-MSC; 3 doses (10×10^6, 50×10^6, and 100×10^6 AD-MSCs)	Nil	WOMAC, KSS, KOOS, VAS, MRI, size and depth of the cartilage defect, the signal intensity of regenerated cartilage and cartilage volume	No TEAE; WOMAC, KSS, and KOOS, improved knee function; VAS, improved pain (statistical significance in high dose); MRI, improvements in all parameters	24 months	[101]
Garay Mendoza et al. (2017)	61	Open-label, phase I/phase II controlled trial	Knee OA	Cell group, BM stimulation with subcutaneous administration of G-CSF ($n = 30$)	Control group, oral acetaminophen ($n = 31$)	VAS and WOMAC scores	BM-SC group showed significant improvement in knee pain and quality of life	6 months	[102]
Kuah et al. (2018)	20	RCT, double-blind, placebo-controlled	KL grade 1 to grade 3	Randomized 4:1; Progenza (PRG) (Allo AD-MSC + culture supernatant); 2 groups, 8 pts. each, 3.9 or 6.7 million cells	4 patients, placebo administered	Safety, WOMAC, VAS, AQoL-4D, biomarkers (urine, C2C and CTX-II; serum, MIF and CTX-I; MRI, MOAKS score)	All patients experienced at least one TEAE; VAS and WOMAC, statistically significant within-group reduction from baseline in PRG group, no statistically significant differences at any time point between placebo and PRG groups; MRI, no decrease in lateral tibial cartilage volume while the placebo group showed a statistically significant cartilage loss	12 months	[103]

Author/ year	Sample size	Study design	Grade of OA	Cell type and dose	Control group	Outcome measures	Outcomes	Follow-up period	Reference
Matas et al. (2018)	29	Phase I/phase II RCT, triple-blind trial	KL grade 1 to grade 3	Allo UC-MSC, single (20×10^6) or repeat dose (20×10^6 baseline and 6 months), 10 pts. each	9 patients, HA (baseline and 6 months)	VAS, WOMAC, MRI, WORMS score	No SAEs, repeat dose group had a significant decrease in VAS and WOMAC scores as compared to HA group, no changes in function subscale, SF36, and MRI	12 months	[48]
Emadedin et al. (2018)	43	RCT, phase I/phase II, placebo-controlled, triple-blind	KL grade 2 to grade 4	Auto BMMSC, 40×10^6 cells (n = 19)	5 ml normal saline (placebo) (n = 24)	VAS, WOMAC, walking distance, painless walking distance, standing time and knee flexion compared	WOMAC, significant improvements in total score, pain, and physical function subscales and improvement in painless walking distance compared with placebo	6 months	[104]
Khalifeh Soltani et al. (2019)	20	RCT, double-blind, placebo-controlled	KL grade 2 to grade 4	Placental-derived MSC, 50–60 x 10^6 cells (n = 10)	Normal saline (n = 10)	VAS, KOOS, knee flexion range of motion (ROM), MRI	No SAEs; significant knee ROM improvement at 2 and 24 weeks; VAS, no change; KOOS, improvement till 8 weeks; MRI, chondral thickness improved in about 10% of the total knee joint area AT 24 weeks	24 weeks	[105]

AAPBSC, autologous activated peripheral blood stem cells; AD-MSCs, adipose tissue-derived mesenchymal stromal cells; AEs, adverse events; Auto, autologous; Allo, allogeneic; AM, arthroscopic microfracture; AQoL-4D, assessment of quality of life 4D questionnaire; BMAC, bone marrow aspirate concentrate; BMMSCs, bone marrow-derived mesenchymal stromal cells; C2C, type II collagen C2C peptide; CTX-1, C-terminal telopeptide of type I collagen; CTX-II, C-terminal telopeptide of type II collagen; dGEMRIC, delayed gadolinium-enhanced magnetic resonance imaging of cartilage; FRI, functional rating index; HA, hyaluronic acid; G-CSF, granulocyte colony-stimulating factor; IA, intra-articular; ICOAP, Intermittent and Constant Osteoarthritis Pain; ICRS, International Cartilage Repair Society Cartilage Injury Evaluation Package; IKDC, International Knee Documentation Committee Subjective Knee Evaluation Form; JKOM, Japanese Knee Osteoarthritis Measure; KL grade, Kellgren and Lawrence grade; KOOS, Knee Injury and Osteoarthritis Outcome Scores; KSS, Knee Society clinical rating system; MIF, macrophage migration inhibitory factor; MOAKS, MRI Osteoarthritis Knee Score; MOCART, magnetic resonance observation of cartilage repair tissue; MCS, mesenchymal cell stimulation; MRI, magnetic resonance imaging; NC, nucleated cells; PCI, poor cartilage index; PBPC, peripheral blood progenitor cells; PGA, patient global assessment; PR-FG, platelet-rich fibrin glue; PRP, platelet-rich plasma; QOL, quality of life; RCT, randomized controlled trial; RHSSK, Revised Hospital for Special Surgery knee scores; ROM, range of motion; SAE, serious adverse event; SAS, Short Arthritis Assessment Scale; SF-36, Short Form-36 quality of life questionnaire; SUSAR, suspected unexpected serious adverse reaction; SVF, stromal vascular fraction; TEAE, treatment emergent adverse event; TKA, total knee arthroplasty; TUG, timed up and go; UC-MSC, umbilical cord-derived MSC; VAS, visual analog pain score; WOMAC, Western Ontario and McMaster Universities Osteoarthritis Index; WORMS, whole-organ magnetic resonance imaging score.

Table 2.
Chronological list of publications of stem cell application for cartilage repair.

b. **Best source of MSC for treatment of OA**: Many studies have been published using different sources of MSCs, and there is no consensus as to which MSC type is the most effective in treating OA. Recently few studies have been published using SVF, bone marrow aspirate concentrate, and micro-fragmented adipose tissue, which further adds to the variability of this issue. The most common problem affecting the clinical outcome in OA is the tendency of MSCs to differentiate into fibrous-like tissue instead of hyaline cartilage [108]. To eliminate or reduce chondrogenesis of the injected MSCs, one school of thought is to identify new sources of MSCs for cartilage repair. Recently synovium-derived stem cells have been used for OA study as it is believed that epigenetic memory may play a role and impact the specific lineage differentiation of MSCs [109]. Hence, the use of synovium stem cells predicts a better outcome as chondrogenic differentiation is expected as it belongs to the same lineage. Fetal stem cells have higher plasticity and proliferation ability than adult stem cells. Hence, fetal tissue-derived stem cells, especially derived from the fetal cartilage, may show higher chondrogenic activity [110] and may be the ideal source of cells for OA. More controlled clinical trials are required to come to a conclusion as to which cell type may be the best choice for the effective treatment of OA.

c. **Autologous or allogeneic source of MSCs**: Most of the published trials used autologous MSCs to minimize immune response, which may lead to best clinical outcomes. Six of the studies in **Table 2** attempted to investigate the potential application of allogeneic MSCs [43, 48, 93, 96, 103, 105] in OA. Recently in the last 2–3 years, most of the studies have attempted to use an allogeneic source of MSCs due to the ease of application. Further, no observed serious adverse effects indicate the safety of allogeneic cells in OA. Around 3000 patients have been administered allogeneic MSCs for different conditions, and no immune response has been reported to date [111]. In a recently published trial using allogeneic umbilical cord-derived MSCs (UC-MSC) in knee OA, patients were randomized to receive hyaluronic acid at baseline and 6 months (HA, $n = 8$), single-dose (20×10^6) UC-MSC at baseline (MSC-1, $n = 9$), or repeated UC-MSC doses at baseline and 6 months ($20 \times 10^6 \times 2$; MSC-2, n = 9). No serious adverse events were reported. At 12 months of follow-up, MSC-2-treated group had significantly lower levels of pain [visual analog score (VAS), Western Ontario and McMaster Universities Osteoarthritis Index (WOMAC), total score, and pain subscale] than HA group [48]. Hence, it can be safely concluded that the use of allogeneic MSCs is safe and may be efficacious in OA.

d. **The optimal dose of MSCs for best efficacy in OA**: MSCs have been used in different doses in several clinical trials of OA (**Table 2**). The dose varied from as low as 1.18 million cells [79] to as high as 150 million cells [43]. In a study by Koh et al. [79], 18 patients were given intra-articular injections with adipose tissue-derived MSCs in a mean dose of 1.18 million cells and platelet-rich plasma. At 26 months of follow-up, patients had significant improvement in VAS, Lysholm, and WOMAC scores. Magnetic resonance imaging (MRI) was evaluated using WORMS score and showed statistically significant improvement in the total and cartilage scores. In another dose-finding study, Pers et al. [90] recruited 18 patients who were treated with autologous AD-MSCs in three different doses: low dose (2×10^6 cells), medium dose (10×10^6 cells), and high dose (50×10^6 cells). After 6 months of follow-up, the procedure was found to be safe, and no serious adverse events were reported. Patients in the low dose had significant improvement in pain levels and functions as compared to baseline. In a dose-finding study conducted by Gupta et al. [43], four different doses (25, 50, 75 and 150 million cells) of allogeneic BMMSCs were used in a total of 60 patients. At 1 year of follow-up, the lower doses of 25 million had shown improvement in pain levels and function as compared to placebo and baseline. However in a study by Jo et al. [101], 18 patients were injected with autologous AD-MSCs in three different doses: 10, 50, and 100 million cells. At 2 years of follow-up, significant improvement in the Knee Society clinical rating system (KSS), Knee Injury and Osteoarthritis Outcome Score (KOOS), and VAS scores was seen in the highest dose of 100 million cells. As can be seen, most of the studies are single-arm studies without any control arm. Hence, to determine the most efficacious dose in OA, more randomized controlled, dose-finding clinical trials are required.

e. **Selection of endpoints for the conduct of clinical trial**: The FDA 2018 draft guidance document for OA regarding the development of structural endpoints for the development of drugs, devices, and biological products for treatment states that approvals for OA to date have been based on patient-reported outcome measures that assess pain and function. For the development of new product in OA, the goal of treatment should be inhibition of structural damage or targeting the underlying pathophysiology associated with OA or significantly delay the complications of joint failure and the need for joint replacement and also to reduce the deterioration of function and worsening of pain. All of the above may be taken into consideration for the development of endpoints for the study in OA [112].

Recently a meta-analysis was done to evaluate the different endpoints used to see the therapeutic efficacy and safety of MSCs for the treatment of patients with knee osteoarthritis [113]. Five hundred eighty-two patients in 11 randomized controlled trials were included in this meta-analysis. It showed that MSC treatment significantly improved VAS and International Knee Documentation Committee (IKDC) scores after 24 months of follow-up compared to controls. MSC therapy also showed significant improvement in the Western Ontario and McMaster Universities Osteoarthritis Index (WOMAC), Lequesne algofunctional indices (Lequesne), Lysholm knee scale (Lysholm), and Tegner activity scale (Tegner) at 12 or 24 months of follow-up. Hence, all the endpoints used currently for evaluation of efficacy in OA have shown significant improvement in different clinical trials:

f. **MRI to evaluate cartilage regeneration**: MRI has emerged as the leading method of imaging soft tissue structures around joints. An ideal MRI study for the cartilage should provide an accurate assessment of cartilage thickness and volume, show morphologic changes of the cartilage surface, show internal cartilage signal changes, and allow evaluation of the subchondral bone for signal abnormalities. Also, it would be desirable for MRI to provide an evaluation of the underlying cartilage physiology, including providing information about the status of the glycosaminoglycan (GAG) and collagen matrices [114]. But, in actual, there is an absence of a standard system by MRI to evaluate cartilage regeneration. Many studies as given in **Table 2** that have used MRI to evaluate cartilage regeneration are only qualitative. It is recommended to use validated imaging outcomes for cartilage regeneration for scientifically validating cell-based therapies, thus advancing the field. The most common parameters used for evaluation of cartilage regeneration by MRI are cartilage thickness in different points in all the compartments of the joint [97], cartilage volume [101], whole-organ magnetic resonance imaging score (WORMS) [43, 48], T2 relaxation time mapping [78, 83, 85, 95, 98], MRI Osteoarthritis Knee Score (MOAKS) score [88, 103], magnetic resonance observation of cartilage repair tissue (MOCART) score [88], and contrast-enhanced imaging technique known as delayed gadolinium-enhanced MRI of cartilage (dGEMRIC) [90]. Among all the parameters, T2 mapping and WORMS seem to be the most commonly used qualitative parameters used for evaluation of cartilage regeneration as it is sensitive to both changes in cartilage hydration and collagen fibril orientation. In a study by Orozco et al. [78], T2 relaxation measurements demonstrated a highly significant decrease of poor cartilage areas (on average, 27%), with the improvement of cartilage quality in 11 of the 12 patients. In another study by Rich et al. [83], a total of 50 patients was evaluated by T2 mapping at 12 months of follow-up after administration of autologous BMMSCs. The mean poor cartilage index (PCI) significantly decreased in 37 of 50 patients (74%), 10 remained the same (20%), and 3 worsened between 7 and 10% (6%). Hence, cartilage T2 mapping may be a sensitive marker for monitoring cartilage quality in subjects with knee OA as it allows us to accurately determine the grade of disorganization of the extracellular matrix.

g. **Use of MSC alone or MSC with a scaffold for intra-articular injection in OA**: When MSCs are injected intra-articularly alone, MSCs scatter widely in the joint, making it impossible to obtain consistent local concentration at the site of cartilage defect. Hence, with a hope to enhance their efficacy in cartilage regeneration, MSC implantation using scaffolds is being attempted in different clinical trials so that the cells are delivered to the site of interest. Compared to direct intra-articular injection, MSC delivery via a scaffold affords more control of proliferation, matrix production, and self-renewal which may help in the regeneration/repair of degenerated or damaged articular cartilage. Different scaffolds have been designed as the delivery system for the repair of articular cartilage. The different scaffolds which can be used are either made of poly-lactic-co-glycolic acids (PLGA) [115], collagen [116], gelatin [117], tricalcium (TCP) [118], poly-lactic acid (PLA) [115], hyaluronic acid (HA) [119], poly glycolic acid (PGA), or fibrin glue [120]. HA has been used frequently for implantation of MSCs into the joint. Many clinical studies (**Table 2**) have used HA as scaffold along with MSCs for implantation of the cells. Cartistem®, an approved drug by the Korean FDA for knee OA, is a combination of human umbilical cord blood-derived MSCs and sodium hyaluronate which is directly implanted at the site of cartilage injury into the joint by arthroscopy [96, 121]. Hence, cells with scaffold are the ideal combination for intra-articular delivery for cartilage degeneration. However, further studies are necessary to find optimal implantation vehicles that can result in the regeneration of articular cartilage.

8.3 Clinical trials in India

Few clinical trials using autologous or allogeneic MSCs or mononuclear stem cells in OA have been conducted in India. The trials registered in the Clinical Trials Registry of India are the two trials done by Stempeutics (one phase II trial completed and the other phase III trial ongoing). However, one published trial by Bansal et al. [122] for the single-arm study was done in India in which a total of 10 patients were treated with AD-MSCs. The patients were evaluated for safety, WOMAC, 6-minute walk test (6MWT), and MRI for cartilage thickness. The patients were followed up for 2 years. The total WOMAC and its subscale scores and 6MWT were significantly improved at all-time points till 2 years of follow-up. Cartilage thickness as determined by MRI improved by at least 0.2 mm in six patients, was unchanged in two patients, and decreased by at least 0.2 mm in two patients. The authors concluded that the procedure demonstrated a strong safety profile with no severe adverse events or complications reported.

8.4 Stempeutics Research experience in osteoarthritis of the knee joint

The off-the-shelf allogeneic, pooled BMMSC product developed by Stempeutics has completed one phase II clinical trial [43] and currently ongoing phase III trial in knee OA. In our completed phase II trial, we included patients of idiopathic OA in grade 2 or 3 of Kellgren and Lawrence radiographic criteria; patients who had self-reported difficulty in at least one of the following activities attributed to knee pain, lifting and carrying groceries, walking 400 m, getting in and out of a chair, or going up and down stairs; and patients who had been on stable medication, including nonsteroidal anti-inflammatory drugs/opioid analgesics for the past 3 months and in the age group of 40–70 years. All the criteria have to be present before being included in the study [43].

8.4.1 Phase II study in patients with osteoarthritis of the knee joint

The phase II results of Stempeucel® in OA patients have been published [43]. Briefly, it was a double-blind, randomized, placebo-controlled, dose-finding study. In this study, 60 OA patients were randomized to receive different doses of Stempeucel®, 25, 50, 75, and 150 million cells or placebo. Stempeucel® was administered intra-articularly (IA) to the knee joint followed by 2 ml of hyaluronic acid (20 mg). The subjects were followed up for 2 years and were evaluated for safety parameters including AEs, and for efficacy parameters, VAS for pain, Intermittent and Constant Osteoarthritis Pain (ICOAP), WOMAC (total score and its subscales), and MRI were done to evaluate the WORMS score. The intra-articular administration of Stempeucel® was safe with knee pain and swelling as the most common AEs. Clinically relevant improvement in a persistent manner was seen in 25 million dose group in all subjective parameters (VAS, ICOAP, and WOMAC scores) (**Figures 2–4**). WORMS of MRI knee did not reveal any difference from the baseline and placebo group. It was concluded that intra-articular administration of Stempeucel® is safe and 25 million dose may be the most effective among the doses tested.

Currently, we are conducting a phase III trial in OA of the knee joint. This is a randomized, double-blind, multicentric, placebo-controlled study assessing the efficacy and safety of intra-articular administration of Stempeucel® in patients with osteoarthritis of the knee joint. One hundred and forty-six patients will be

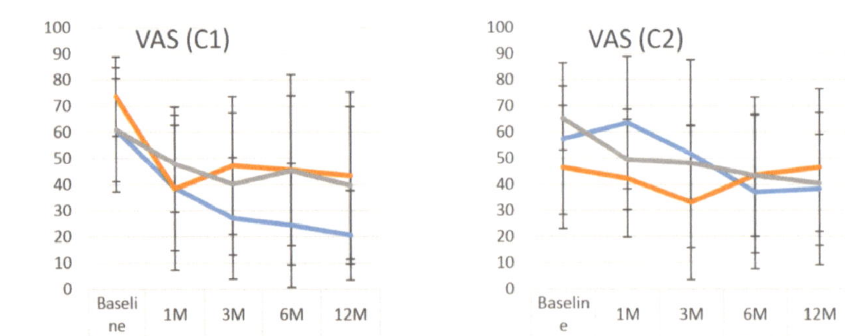

	Baseline	1M	3M	6M	12M
25M	60.9	38.7	27.1	24.4	20.6
50M	73.7	38.4	47.3	45.6	43.4
P1	61	48	40.2	45.3	39.7

	Baseline	1M	3M	6M	12M
75M	57.4	63.6	51.7	37.1	38.3
150M	46.6	42.3	33.1	43.6	46.6
P2	65.3	49.5	48.3	43.4	40.5

Figure 2.
Visual analog scale values. Data presented as mean value ± SD; C1 = cohort 1; C2 = cohort 2; 25M, 50M, 75M, 150M = 25, 50, 75, 150 million cells, respectively; 1M, 3M, 6M, 12M = 1, 3, 6, 12 months, respectively.

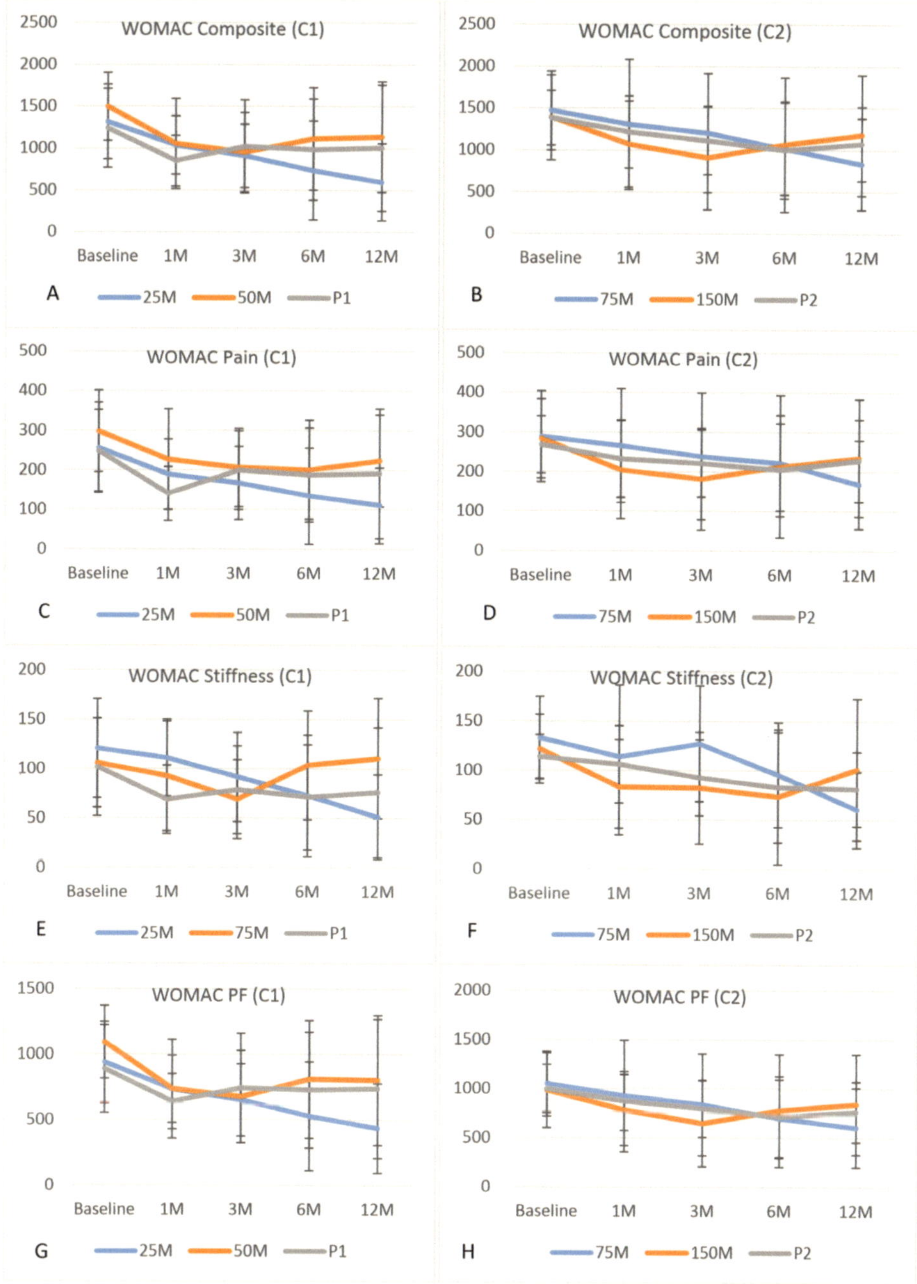

Figure 3.
WOMAC results. WOMAC: (A, B) composite; (C, D) pain; (E, F) stiffness; and (G, H) physical function. Data presented as mean value ± SD; C1 = cohort 1; C2 = cohort 2; 25M, 50M, 75M, 150M = 25, 50, 75, 150 million cells, respectively; 1M, 3M, 6M, 12M = 1, 3, 6, 12 months, respectively; WOMAC = Western Ontario and McMaster Universities Osteoarthritis Index.

randomized to stem cell and placebo arm in a ratio of 1:1. Seventy-three patients will receive Stempeucel® (25 million) followed by 2 ml of hyaluronan, and 73 patients will receive only intra-articular injection of 2 ml of placebo followed by 2 ml of hyaluronan. The patients will be followed up for a total of 2 years after IMP administration. The details of the study are found in the Clinical Trials Registry of India (CTRI/2018/09/015785).

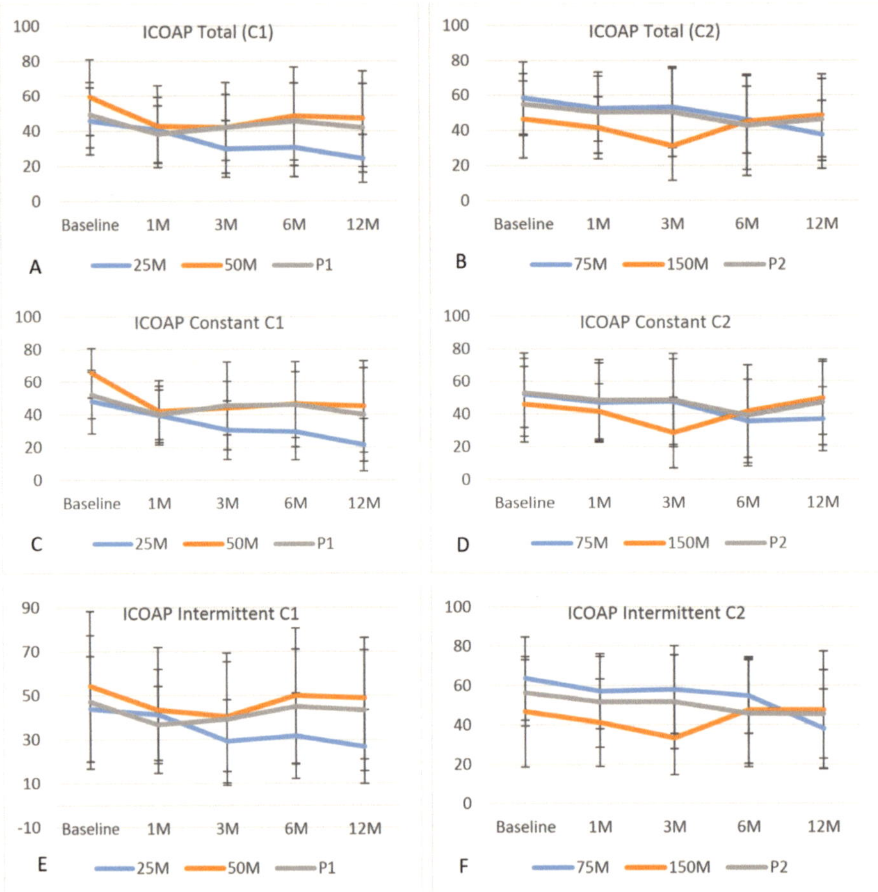

Figure 4.
ICOAP results. ICOAP: (A, B) total; (C, D) continuous pain; and (E, F) intermittent pain. Data presented as mean value ± SD; C1 = cohort 1; C2 = cohort 2; 25M, 50M, 75M, 150M = 25, 50, 75, 150 million cells, respectively; 1M, 3M, 6M, 12M = 1, 3, 6, 12 months, respectively; ICOAP = Intermittent and Constant Osteoarthritis Pain.

9. Conclusion

Osteoarthritis is a common disorder involving damage to synovial joint tissues particularly the cartilage and bone. Current treatments are mostly targeted at end-stage disease, but biological therapies including stem cell therapy show a promise for earlier intervention with a more prolonged benefit. With all the published clinical trial data, it is reasonable to expect that MSCs may prove to be an important therapy for OA. Pooled BMMSCs with their enhanced anti-inflammatory potential, immunomodulatory properties, and secretion of paracrine factors create the optimum environment for a controlled reparative pathway in the affected joint. Pooled BMMSC treatment, perhaps combined with other modalities like a scaffold, would be advantageous in providing treatment in early OA to slow disease progression, thus delaying or avoiding total joint replacement.

Acknowledgements

We thank Stempeutics Research Pvt. Ltd. and Mr. BN Manohar for proving the required funding for the relevant studies. We would like to thank

Dr. Anish S Majumdar and Dr. Sudha Balasubramanian for their critical scientific inputs for potency assay development, preclinical studies, and clinical studies conducted by Stempeutics Research. We also thank Dr. Anoop Chullikana for his valuable inputs in the clinical studies.

Conflict of interest

The authors declare no conflict of interests.

Author details

Charan Thej and Pawan Kumar Gupta*
Stempeutics Research Pvt. Ltd., Manipal Hospitals Whitefield Pvt. Ltd., Bengaluru, India

*Address all correspondence to: pawan.gupta@stempeutics.com

References

[1] Yucesoy B, Charles LE, Baker B, Burchfiel CM. Occupational and genetic risk factors for osteoarthritis: A review. Work [Internet]. 2015;**50**(2):261-273. Available from: http://www.ncbi.nlm.nih.gov/pubmed/24004806

[2] Hame SL, Alexander RA. Knee osteoarthritis in women. Current Reviews in Musculoskeletal Medicine [Internet]. 2013;**6**(2):182-187. Available from: http://www.ncbi.nlm.nih.gov/pubmed/23471773

[3] Pal CP, Singh P, Chaturvedi S, Pruthi KK, Vij A. Epidemiology of knee osteoarthritis in India and related factors. Indian Journal of Orthopaedics [Internet]. 2016;**50**(5):518-522. Available from: http://www.ncbi.nlm.nih.gov/pubmed/27746495

[4] Musumeci G, Aiello FC, Szychlinska MA, Di Rosa M, Castrogiovanni P, Mobasheri A. Osteoarthritis in the XXIst century: Risk factors and behaviours that influence disease onset and progression. International Journal of Molecular Sciences [Internet]. 2015;**16**(3):6093-6112. Available from: http://www.ncbi.nlm.nih.gov/pubmed/25785564

[5] Martel-Pelletier J, Barr AJ, Cicuttini FM, Conaghan PG, Cooper C, Goldring MB, et al. Osteoarthritis. Nature Reviews Disease Primers [Internet]. 2016;**2**:16072. Available from: http://www.nature.com/articles/nrdp201672

[6] Stubendorff JJ, Lammentausta E, Struglics A, Lindberg L, Heinegård D, Dahlberg LE. Is cartilage sGAG content related to early changes in cartilage disease? Implications for interpretation of dGEMRIC. Osteoarthritis and Cartilage [Internet]. 2012;**20**(5):396-404. Available from: https://linkinghub.elsevier.com/retrieve/pii/S1063458412000398

[7] Hermann W, Lambova S, Müller-Ladner U. Current treatment options for osteoarthritis. Current Rheumatology Reviews [Internet]. 2018;**14**(2):108-116. Available from: http://www.eurekaselect.com/155279/article

[8] Chang DS, Hsu E, Hottinger DG, Cohen SP. Anti-nerve growth factor in pain management: Current evidence. Journal of Pain Research [Internet]. 2016;**9**:373-383. Available from: http://www.ncbi.nlm.nih.gov/pubmed/27354823

[9] Vasiliadis HS, Tsikopoulos K. Glucosamine and chondroitin for the treatment of osteoarthritis. World Journal of Orthopedics [Internet]. 2017;**8**(1):1-11. Available from: http://www.ncbi.nlm.nih.gov/pubmed/28144573

[10] Kohn MD, Sassoon AA, Fernando ND. Classifications in brief: Kellgren-Lawrence classification of osteoarthritis. Clinical Orthopaedics and Related Research [Internet]. 2016;**474**(8):1886-1893. Available from: http://www.ncbi.nlm.nih.gov/pubmed/26872913

[11] Zellner J, Krutsch W, Pfeifer C, Koch M, Nerlich M, Angele P. Autologous chondrocyte implantation for cartilage repair: Current perspectives. Orthopedic Research and Reviews [Internet]. 11 Nov 2015;**7**:149-158. Available from: https://www.dovepress.com/autologous-chondrocyte-implantation-for-cartilage-repair-current-persp-peer-reviewed-article-ORR

[12] Dhillon MS, Patel S, John R. PRP in OA knee—Update, current confusions and future options. SICOT-J [Internet]. 2017;**3**:27. Available from: http://www.ncbi.nlm.nih.gov/pubmed/28322719

[13] Jungbluth P, Grassmann J-P, Thelen S, Wild M, Sager M, Windolf J, et al. Concentration of platelets and growth factors in platelet-rich plasma from Goettingen minipigs. GMS Interdisciplinary Plastic and Reconstructive Surgery DGPW.

2014;**3**:Doc11. Available from: http://www.ncbi.nlm.nih.gov/pubmed/26504722

[14] Glynn LG, Mustafa A, Casey M, Krawczyk J, Blom J, Galvin R, et al. Platelet-rich plasma (PRP) therapy for knee arthritis: A feasibility study in primary care. Pilot and Feasibility Studies [Internet]. 2018;**4**:93. Available from: http://www.ncbi.nlm.nih.gov/pubmed/29997901

[15] Friedenstein AJ. Precursor cells of mechanocytes. International Review of Cytology [Internet]. 1976;**47**:327-359. Available from: http://www.ncbi.nlm.nih.gov/pubmed/11195

[16] Owen M, Friedenstein AJ. Stromal stem cells: Marrow-derived osteogenic precursors. Ciba Foundation Symposium [Internet]. 1988;**136**:42-60. Available from: http://www.ncbi.nlm.nih.gov/pubmed/3068016

[17] Pittenger MF, Mackay AM, Beck SC, Jaiswal RK, Douglas R, Mosca JD, et al. Multilineage potential of adult human mesenchymal stem cells. Science [Internet]. 1999;**284**(5411):143-147. Available from: http://www.ncbi.nlm.nih.gov/pubmed/10102814

[18] Caplan AI. Mesenchymal stem cells. Journal of Orthopaedic Research [Internet]. 1991;**9**(5):641-650. Available from: http://www.ncbi.nlm.nih.gov/pubmed/1870029

[19] Bianco P, Gehron RP. Marrow stromal stem cells. Journal of Clinical Investigation [Internet]. 2000;**105**(12):1663-1668. Available from: http://www.jci.org/articles/view/10413

[20] Dominici M, Le Blanc K, Mueller I, Slaper-Cortenbach I, Marini F, Krause D, et al. Minimal criteria for defining multipotent mesenchymal stromal cells. The International Society for Cellular Therapy position statement. Cytotherapy [Internet]. 2006;**8**(4):315-317. Available from: http://www.sciencedirect.com/science/article/pii/S1465324906708817

[21] Le Blanc K, Ringdén O. Immunomodulation by mesenchymal stem cells and clinical experience. Journal of Internal Medicine. 2007;**262**(5):509-525

[22] Abdi R, Fiorina P, Adra CN, Atkinson M, Sayegh MH. Immunomodulation by mesenchymal stem cells: A potential therapeutic strategy for type 1 diabetes. Diabetes. 2008;**57**(7):1759-1767

[23] Mattar P, Bieback K. Comparing the immunomodulatory properties of bone marrow, adipose tissue, and birth-associated tissue mesenchymal stromal cells. Frontiers in Immunology. 2015;**6**(Nov):1-8

[24] Bijlsma JW, Berenbaum F, Lafeber FP. Osteoarthritis: An update with relevance for clinical practice. Lancet [Internet]. 2011;**377**(9783):2115-2126. Available from: http://linkinghub.elsevier.com/retrieve/pii/S0140673611602432

[25] Gupta PK, Das AK, Chullikana A, Majumdar AS. Mesenchymal stem cells for cartilage repair in osteoarthritis. Stem Cell Research & Therapy [Internet]. 2012;**3**(4):25. Available from: http://www.ncbi.nlm.nih.gov/pubmed/22776206

[26] Nauta AJ, Fibbe WE. Immunomodulatory properties of mesenchymal stromal cells. Blood. 2007;**110**(10):3499-3506

[27] Liu Q, Zheng H, Chen X, Peng Y, Huang W, Li X, et al. Human mesenchymal stromal cells enhance the immunomodulatory function of CD8(+)CD28(−) regulatory T cells. Cellular & Molecular Immunology [Internet]. Nov 2015;**12**(6):708-718. Available from: http://www.nature.com/doifinder/10.1038/cmi.2014.118%5Cnhttp://www.ncbi.nlm.nih.gov/pubmed/25482073

[28] Freitag J, Bates D, Boyd R, Shah K, Barnard A, Huguenin L, et al. Mesenchymal stem cell therapy in the treatment of osteoarthritis: Reparative pathways, safety and efficacy—A review. BMC Musculoskeletal Disorders [Internet]. 2016;**17**:230. Available from: http://www.ncbi.nlm.nih.gov/pubmed/27229856

[29] Wang M, Shen J, Jin H, Im H-J, Sandy J, Chen D. Recent progress in understanding molecular mechanisms of cartilage degeneration during osteoarthritis. Annals of the New York Academy of Sciences [Internet]. 2011;**1240**:61-69. Available from: http://www.ncbi.nlm.nih.gov/pubmed/22172041

[30] Nauta AJ. Donor-derived mesenchymal stem cells are immunogenic in an allogeneic host and stimulate donor graft rejection in a nonmyeloablative setting. Blood [Internet]. 2006;**108**(6):2114-2120. Available from: http://www.bloodjournal.org/cgi/doi/10.1182/blood-2005-11-011650

[31] Vaes B, Van't Hof W, Deans R, Pinxteren J. Application of MultiStem(®) allogeneic cells for immunomodulatory therapy: Clinical progress and pre-clinical challenges in prophylaxis for graft versus host disease. Frontiers in Immunology [Internet]. 2012;**3**:345. Available from: http://www.ncbi.nlm.nih.gov/pubmed/23205020

[32] Ketterl N, Brachtl G, Schuh C, Bieback K, Schallmoser K, Reinisch A, et al. A robust potency assay highlights significant donor variation of human mesenchymal stem/progenitor cell immune modulatory capacity and extended radio-resistance. Stem Cell Research & Therapy [Internet]. 2015;**6**(1):236. Available from: http://stemcellres.com/content/6/1/236

[33] Phinney DG, Kopen G, Righter W, Webster S, Tremain N, Prockop DJ. Donor variation in the growth properties and osteogenic potential of human marrow stromal cells. Journal of Cellular Biochemistry [Internet]. 1999;**75**(3):424-436. Available from: http://www.ncbi.nlm.nih.gov/pubmed/10536366

[34] Mindaye ST, Ra M, Lo Surdo JL, Bauer SR, Alterman MA. Global proteomic signature of undifferentiated human bone marrow stromal cells: Evidence for donor-to-donor proteome heterogeneity. Stem Cell Research [Internet]. 2013;**11**(2):793-805. Available from: http://linkinghub.elsevier.com/retrieve/pii/S187350611300055X

[35] Le Blanc K, Frassoni F, Ball L, Locatelli F, Roelofs H, Lewis I, et al. Mesenchymal stem cells for treatment of steroid-resistant, severe, acute graft-versus-host disease: A phase II study. Lancet [Internet]. 2008;**371**(9624):1579-1586. Available from: http://linkinghub.elsevier.com/retrieve/pii/S014067360860690X

[36] Martin PJ, Uberti JP, Soiffer RJ, Klingemann H, Waller EK, Daly AS, et al. Prochymal improves response rates in patients with steroid-refractory acute graft versus host disease (SR-GVHD) involving the liver and gut: Results of a randomized, placebo-controlled, multicenter phase III trial in GVHD. Biology of Blood and Marrow Transplantation [Internet]. 2010;**16**(2):S169-S170. Available from: http://linkinghub.elsevier.com/retrieve/pii/S1083879109006533

[37] Galipeau J. The mesenchymal stromal cells dilemma-does a negative phase III trial of random donor mesenchymal stromal cells in steroid-resistant graft-versus-host disease represent a death knell or a bump in the road? Cytotherapy [Internet]. 2013;**15**(1):2-8. DOI: 10.1016/j.jcyt.2012.10.002

[38] Samuelsson H, Ringdén O, Lönnies H, Le Blanc K. Optimizing in vitro

conditions for immunomodulation and expansion of mesenchymal stromal cells. Cytotherapy. 2009;**11**(2):129-136

[39] Kuçi Z, Bönig H, Kreyenberg H, Bunos M, Jauch A, Janssen JWG, et al. Mesenchymal stromal cells from pooled mononuclear cells of multiple bone marrow donors as rescue therapy in pediatric severe steroid-refractory graft-versus-host disease: A multicenter survey. Haematologica [Internet]. 2016;**101**(8):985-994. Available from: http://www.ncbi.nlm.nih.gov/pubmed/27175026

[40] Bader P, Kuçi Z, Bakhtiar S, Basu O, Bug G, Dennis M, et al. Effective treatment of steroid and therapy-refractory acute graft-versus-host disease with a novel mesenchymal stromal cell product (MSC-FFM). Bone Marrow Transplantation [Internet]. Jul 2018;**53**(7):852-862. Available from: http://www.nature.com/articles/s41409-018-0102-z

[41] Mamidi MK, Nathan KG, Singh G, Thrichelvam ST, Mohd Yusof NAN, Fakharuzi NA, et al. Comparative cellular and molecular analyses of pooled bone marrow multipotent mesenchymal stromal cells during continuous passaging and after successive cryopreservation. Journal of Cellular Biochemistry. 2012;**113**(10):3153-3164

[42] Chullikana A, Majumdar AS, Gottipamula S, Krishnamurthy S, Kumar AS, Prakash VS, et al. Randomized, double-blind, phase I/II study of intravenous allogeneic mesenchymal stromal cells in acute myocardial infarction. Cytotherapy. 2015;**17**:250-261

[43] Gupta PK, Chullikana A, Rengasamy M, Shetty N, Pandey V, Agarwal V, et al. Efficacy and safety of adult human bone marrow-derived, cultured, pooled, allogeneic mesenchymal stromal cells (Stempeucel®): Preclinical and clinical trial in osteoarthritis of the knee joint. Arthritis Research & Therapy. 2016;**18**(1):301. Available from: http://arthritis-research.biomedcentral.com/articles/10.1186/s13075-016-1195-7

[44] Carmen J, Burger SR, McCaman M, Rowley JA. Developing assays to address identity, potency, purity and safety: cell characterization in cell therapy process development. Regenerative Medicine [Internet]. 2012;**7**(1):85-100. Available from: http://www.ncbi.nlm.nih.gov/pubmed/22168500

[45] Dell'Accio F, Vanlauwe J, Bellemans J, Neys J, De Bari C, Luyten FP. Expanded phenotypically stable chondrocytes persist in the repair tissue and contribute to cartilage matrix formation and structural integration in a goat model of autologous chondrocyte implantation. Journal of Orthopaedic Research [Internet]. 2003;**21**(1):123-131. Available from: http://www.ncbi.nlm.nih.gov/pubmed/12507589

[46] Bravery CA, Carmen J, Fong T, Oprea W, Hoogendoorn KH, Woda J, et al. Potency assay development for cellular therapy products: An ISCT* review of the requirements and experiences in the industry. Cytotherapy [Internet]. 2013;**15**(1):9-19. DOI: 10.1016/j.jcyt.2012.10.008

[47] Jeong SY, Kim DH, Ha J, Jin HJ, Kwon S-J, Chang JW, et al. Thrombospondin-2 secreted by human umbilical cord blood-derived mesenchymal stem cells promotes chondrogenic differentiation. Stem Cells [Internet]. 2013;**31**(10):2136-2148. Available from: http://www.ncbi.nlm.nih.gov/pubmed/23843355

[48] Matas J, Orrego M, Amenabar D, Infante C, Tapia-Limonchi R, Cadiz MI, et al. Umbilical cord-derived mesenchymal stromal cells (MSCs) for knee osteoarthritis: Repeated MSC dosing is superior to a single MSC dose and to hyaluronic acid in a controlled

randomized phase I/II trial. Stem Cells Translational Medicine. Mar 2019;**8**(3):215-224

[49] Thorup A-S, Eldridge SE, Caxaria S, Nalesso G, Thomas BL, Pitzalis C, et al. A potency assay for assessing the chondrogenic efficiency of bioactive molecules in human cartilage in vivo. Osteoarthritis and Cartilage [Internet]. 2017;**25**:S273. Available from: https://linkinghub.elsevier.com/retrieve/pii/S1063458417305113

[50] Wang Y, Han Z, Song Y, Han ZC. Safety of mesenchymal stem cells for clinical application. Stem Cells International [Internet]. 2012;**2012**:1-4. Available from: http://www.hindawi.com/journals/sci/2012/652034/

[51] Rengasamy M, Gupta PK, Kolkundkar U, Singh G, Balasubramanian S, SundarRaj S, et al. Preclinical safety & toxicity evaluation of pooled, allogeneic human bone marrow-derived mesenchymal stromal cells. The Indian Journal of Medical Research [Internet]. 2016;**144**(6): 852-864. Available from: http://www.ncbi.nlm.nih.gov/pubmed/28474622

[52] Diekman BO, Wu C-L, Louer CR, Furman BD, Huebner JL, Kraus VB, et al. Intra-articular delivery of purified mesenchymal stem cells from C57BL/6 or MRL/MpJ superhealer mice prevents posttraumatic arthritis. Cell Transplantation [Internet]. 2013;**22**(8):1395-1408. Available from: http://www.ncbi.nlm.nih.gov/pubmed/22889498

[53] Kim JE, Lee SM, Kim SH, Tatman P, Gee AO, Kim D-H, et al. Effect of self-assembled peptide-mesenchymal stem cell complex on the progression of osteoarthritis in a rat model. International Journal of Nanomedicine [Internet]. 2014;**9**(Suppl 1):141-157. Available from: http://www.ncbi.nlm.nih.gov/pubmed/24872709

[54] Yang X, Zhu T-Y, Wen L-C, Cao Y-P, Liu C, Cui Y-P, et al. Intraarticular injection of allogenic mesenchymal stem cells has a protective role for the osteoarthritis. Chinese Medical Journal (English) [Internet]. 2015;**128**(18):2516-2523. Available from: http://www.ncbi.nlm.nih.gov/pubmed/26365972

[55] Singh A, Goel SC, Gupta KK, Kumar M, Arun GR, Patil H, et al. The role of stem cells in osteoarthritis: An experimental study in rabbits. Bone & Joint Research [Internet]. 2014;**3**(2): 32-37. Available from: http://www.ncbi.nlm.nih.gov/pubmed/24526748

[56] Chiang E-R, Ma H-L, Wang J-P, Liu C-L, Chen T-H, Hung S-C. allogeneic mesenchymal stem cells in combination with hyaluronic acid for the treatment of osteoarthritis in rabbits. PLoS One [Internet]. 2016;**11**(2):e0149835. Available from: http://www.ncbi.nlm.nih.gov/pubmed/26915044

[57] Xing D, Kwong J, Yang Z, Hou Y, Zhang W, Ma B, et al. Intra-articular injection of mesenchymal stem cells in treating knee osteoarthritis: A systematic review of animal studies. Osteoarthritis and Cartilage [Internet]. 2018;**26**(4):445-461. Available from: https://linkinghub.elsevier.com/retrieve/pii/S1063458418300487

[58] Fahey JR, Katoh H, Malcolm R, Perez AV. The case for genetic monitoring of mice and rats used in biomedical research. Mammalian Genome [Internet]. 2013;**24**(3-4): 89-94. Available from: http://www.ncbi.nlm.nih.gov/pubmed/23314661

[59] Wang X, Omar O, Vazirisani F, Thomsen P, Ekström K. Mesenchymal stem cell-derived exosomes have altered microRNA profiles and induce osteogenic differentiation depending on the stage of differentiation. PLoS One [Internet]. 2018;**13**(2):e0193059. DOI: 10.1371/journal.pone.0193059. Shi X-M, editor

[60] Lo Monaco M, Merckx G, Ratajczak J, Gervois P, Hilkens P, Clegg P, et al. Stem cells for cartilage repair: Preclinical studies and insights in translational animal models and outcome measures. Stem Cells International [Internet]. 2018;**2018**:9079538. Available from: http://www.ncbi.nlm.nih.gov/pubmed/29535784

[61] Frisbie DD, Kisiday JD, Kawcak CE, Werpy NM, McIlwraith CW. Evaluation of adipose-derived stromal vascular fraction or bone marrow-derived mesenchymal stem cells for treatment of osteoarthritis. Journal of Orthopaedic Research [Internet]. 2009;**27**(12):1675-1680. Available from: http://www.ncbi.nlm.nih.gov/pubmed/19544397

[62] McCarthy D, Pandey S, Cekanova M, Biskup J, Drum M, Millis D. The effects of autologous mesenchymal stem cell therapy on elbow osteoarthritis in dogs: A pilot study. Veterinary and Comparative Orthopaedics and Traumatology [Internet]. 2018;**31**(S 02):A1-A25. Available from: http://www.thieme-connect.de/DOI/DOI?10.1055/s-0038-1668216

[63] Song F, Tang J, Geng R, Hu H, Zhu C, Cui W, et al. Comparison of the efficacy of bone marrow mononuclear cells and bone mesenchymal stem cells in the treatment of osteoarthritis in a sheep model. International Journal of Clinical and Experimental Pathology [Internet]. 2014;**7**(4):1415-1426. Available from: http://www.ncbi.nlm.nih.gov/pubmed/24817937

[64] Delling U, Brehm W, Ludewig E, Winter K, Jülke H. Longitudinal evaluation of effects of intra-articular mesenchymal stromal cell administration for the treatment of osteoarthritis in an ovine model. Cell Transplantation [Internet]. 2015;**24**(11):2391-2407. Available from: http://www.ncbi.nlm.nih.gov/pubmed/25581789

[65] Murphy JM, Fink DJ, Hunziker EB, Barry FP. Stem cell therapy in a caprine model of osteoarthritis. Arthritis and Rheumatism [Internet]. 2003;**48**(12):3464-3474. Available from: http://www.ncbi.nlm.nih.gov/pubmed/14673997

[66] Sato M, Uchida K, Nakajima H, Miyazaki T, Guerrero AR, Watanabe S, et al. Direct transplantation of mesenchymal stem cells into the knee joints of Hartley strain guinea pigs with spontaneous osteoarthritis. Arthritis Research & Therapy [Internet]. 2012;**14**(1):R31. Available from: http://www.ncbi.nlm.nih.gov/pubmed/22314040

[67] Suhaeb AM, Naveen S, Mansor A, Kamarul T. Hyaluronic acid with or without bone marrow derived-mesenchymal stem cells improves osteoarthritic knee changes in rat model: a preliminary report. Indian Journal of Experimental Biology [Internet]. 2012;**50**(6):383-390. Available from: http://www.ncbi.nlm.nih.gov/pubmed/22734248

[68] Lalu MM, McIntyre L, Pugliese C, Fergusson D, Winston BW, Marshall JC, et al. Safety of cell therapy with mesenchymal stromal cells (SafeCell): A systematic review and meta-analysis of clinical trials. PLoS One [Internet]. 2012;**7**(10):e47559. Available from: http://www.ncbi.nlm.nih.gov/pubmed/23133515

[69] Peeters CMM, Leijs MJC, Reijman M, van Osch GJVM, Bos PK. Safety of intra-articular cell-therapy with culture-expanded stem cells in humans: A systematic literature review. Osteoarthritis and Cartilage. 2013;**21**(10):1465-1473

[70] Centeno CJ, Al-Sayegh H, Freeman MD, Smith J, Murrell WD, Bubnov R. A multi-center analysis of adverse events among two thousand,

three hundred and seventy two adult patients undergoing adult autologous stem cell therapy for orthopaedic conditions. International Orthopaedics. 2016;**40**(8):1755-1765

[71] Wakitani S, Imoto K, Yamamoto T, Saito M, Murata N, Yoneda M. Human autologous culture expanded bone marrow mesenchymal cell transplantation for repair of cartilage defects in osteoarthritic knees. Osteoarthritis and Cartilage. 2002;**10**(3):199-206

[72] Centeno CJ, Busse D, Kisiday J, Keohan C, Freeman M, Karli D. Increased knee cartilage volume in degenerative joint disease using percutaneously implanted, autologous mesenchymal stem cells. Pain Physician. 2008;**11**(3):343-353

[73] Haleem AM, Singergy AA, Sabry D, Atta HM, Rashed LA, Chu CR, et al. The clinical use of human culture-expanded autologous bone marrow mesenchymal stem cells transplanted on platelet-rich fibrin glue in the treatment of articular cartilage defects: A pilot study and preliminary results. Cartilage. 2010;**1**(4):253-261

[74] Nejadnik H, Hui JH, Feng Choong EP, Tai B-C, Lee EH. Autologous bone marrow-derived mesenchymal stem cells versus autologous chondrocyte implantation: An observational cohort study. The American Journal of Sports Medicine. 2010;**38**(6):1110-1116

[75] Davatchi F, Abdollahi BS, Mohyeddin M, Shahram F, Nikbin B. Mesenchymal stem cell therapy for knee osteoarthritis. Preliminary report of four patients. International Journal of Rheumatic Diseases. 2011;**14**(2):211-215

[76] Saw K-Y, Anz A, Merican S, Tay Y-G, Ragavanaidu K, Jee CSY, et al. Articular cartilage regeneration with autologous peripheral blood progenitor cells and hyaluronic acid after arthroscopic subchondral drilling: A report of 5 cases with histology. Arthroscopy. 2011;**27**(4):493-506

[77] Koh Y-G, Choi Y-J. Infrapatellar fat pad-derived mesenchymal stem cell therapy for knee osteoarthritis. The Knee. 2012;**19**(6):902-907

[78] Orozco L, Munar A, Soler R, Alberca M, Soler F, Huguet M, et al. Treatment of knee osteoarthritis with autologous mesenchymal stem cells: A pilot study. Transplantation. 2013;**95**(12):1535-1541

[79] Koh Y-G, Jo S-B, Kwon O-R, Suh D-S, Lee S-W, Park S-H, et al. Mesenchymal stem cell injections improve symptoms of knee osteoarthritis. Arthroscopy. 2013;**29**(4):748-755

[80] Van Pham P, Bui KH-T, Duong TD, Nguyen NT, Nguyen TD, Le VT, et al. Symptomatic knee osteoarthritis treatment using autologous adipose derived stem cells and platelet-rich plasma: A clinical study. Biomedical Research and Therapy. 2014;**1**(1):2

[81] Koh YG, Choi YJ, Kwon OR, Kim YS. Second-look arthroscopic evaluation of cartilage lesions after mesenchymal stem cell implantation in osteoarthritic knees. The American Journal of Sports Medicine. 2014;**42**(7):1628-1637

[82] Koh Y-G, Choi Y-J, Kwon S-K, Kim Y-S, Yeo J-E. Clinical results and second-look arthroscopic findings after treatment with adipose-derived stem cells for knee osteoarthritis. Knee Surgery, Sports Traumatology, Arthroscopy. 2015;**23**(5):1308-1316

[83] Soler Rich R, Munar A, Soler Romagosa F, Peirau X, Huguet M, et al. Treatment of Knee Osteoarthritis

with Autologous Expanded Bone Marrow Mesenchymal Stem Cells: 50 Cases Clinical and MRI Results at One Year Follow-Up. Journal of Stem Cell Research and Therapy. 2015;**5**:285

[84] Davatchi F, Sadeghi Abdollahi B, Mohyeddin M, Nikbin B. Mesenchymal stem cell therapy for knee osteoarthritis: 5 years follow-up of three patients. International Journal of Rheumatic Diseases. 2016;**19**(3):219-225

[85] Soler R, Orozco L, Munar A, Huguet M, López R, Vives J, et al. Final results of a phase I-II trial using ex vivo expanded autologous mesenchymal stromal cells for the treatment of osteoarthritis of the knee confirming safety and suggesting cartilage regeneration. The Knee. 2016;**23**(4):647-654

[86] Sampson S, Smith J, Vincent H, Aufiero D, Zall M, Botto-van-Bemden A. Intra-articular bone marrow concentrate injection protocol: Short-term efficacy in osteoarthritis. Regenerative Medicine. 2016;**11**(6):511-520

[87] Fodor PB, Paulseth SG. Adipose derived stromal cell (ADSC) injections for pain management of osteoarthritis in the human knee joint. Aesthetic Surgery Journal. 2016;**36**(2):229-236

[88] Kim YS, Choi YJ, Lee SW, Kwon OR, Suh DS, Heo DB, et al. Assessment of clinical and MRI outcomes after mesenchymal stem cell implantation in patients with knee osteoarthritis: A prospective study. Osteoarthritis and Cartilage. 2016;**24**(2):237-245

[89] Pak J, Lee JH, Park KS, Jeong BC, Lee SH. Regeneration of cartilage in human knee osteoarthritis with autologous adipose tissue-derived stem cells and autologous extracellular matrix. BioResearch Open Access. 2016;**5**(1):192-200

[90] Pers Y-M, Rackwitz L, Ferreira R, Pullig O, Delfour C, Barry F, et al. Adipose mesenchymal stromal cell-based therapy for severe osteoarthritis of the knee: A Phase I dose-escalation trial. Stem Cells Translational Medicine. 2016;**5**(7):847-856

[91] Koh Y-G, Kwon O-R, Kim Y-S, Choi Y-J, Tak D-H. Adipose-derived mesenchymal stem cells with microfracture versus microfracture alone: 2-year follow-up of a prospective randomized trial. Arthroscopy. 2016;**32**(1):97-109

[92] Lamo-Espinosa JM, Mora G, Blanco JF, Granero-Moltó F, Nuñez-Córdoba JM, Sánchez-Echenique C, et al. Intra-articular injection of two different doses of autologous bone marrow mesenchymal stem cells versus hyaluronic acid in the treatment of knee osteoarthritis: Multicenter randomized controlled clinical trial (phase I/II). Journal of Translational Medicine. 2016;**14**(1):246

[93] de Windt TS, Vonk LA, Slaper-Cortenbach ICM, van den Broek MPH, Nizak R, van Rijen MHP, et al. Allogeneic mesenchymal stem cells stimulate cartilage regeneration and are safe for single-stage cartilage repair in humans upon mixture with recycled autologous chondrons. Stem Cells. 2017;**35**(1):256-264

[94] Turajane T, Chaveewanakorn U, Fongsarun W, Aojanepong J, Papadopoulos KI. Avoidance of total knee arthroplasty in early osteoarthritis of the knee with intra-articular implantation of autologous activated peripheral blood stem cells versus hyaluronic acid: A randomized controlled trial with differential effects of growth factor. Stem Cells International. 2017;**2017**:8925132

[95] Shapiro SA, Kazmerchak SE, Heckman MG, Zubair AC, O'Connor MI. A prospective, single-blind, placebo-controlled trial of bone marrow aspirate concentrate for knee osteoarthritis. The

American Journal of Sports Medicine. 2017;**45**(1):82-90

[96] Park Y-B, Ha C-W, Lee C-H, Yoon YC, Park Y-G. Cartilage regeneration in osteoarthritic patients by a composite of allogeneic umbilical cord blood-derived mesenchymal stem cells and hyaluronate hydrogel: Results from a clinical trial for safety and proof-of-concept with 7 years of extended follow-up. Stem Cells Translational Medicine. 2017;**6**(2):613-621

[97] Nguyen PD, Tran TD-X, Nguyen HT-N, Vu HT, Le PT-B, Phan NL-C, et al. Comparative clinical observation of arthroscopic microfracture in the presence and absence of a stromal vascular fraction injection for osteoarthritis. Stem Cells Translational Medicine. 2017;**6**(1):187-195

[98] Pintat J, Silvestre A, Magalon G, Gadeau AP, Pesquer L, Perozziello A, et al. Intra-articular injection of mesenchymal stem cells and platelet-rich plasma to treat patellofemoral osteoarthritis: Preliminary results of a long-term pilot study. Journal of Vascular and Interventional Radiology. 2017;**28**(12):1708-1713

[99] Russo A, Condello V, Madonna V, Guerriero M, Zorzi C. Autologous and micro-fragmented adipose tissue for the treatment of diffuse degenerative knee osteoarthritis. Journal of Experimental Orthopaedics. 2017;**4**(1):33

[100] Yokota N, Yamakawa M, Shirata T, Kimura T, Kaneshima H. Clinical results following intra-articular injection of adipose-derived stromal vascular fraction cells in patients with osteoarthritis of the knee. Regenerative Therapy. 2017;**6**:108-112

[101] Jo CH, Chai JW, Jeong EC, Oh S, Shin JS, Shim H, et al. Intra-articular injection of mesenchymal stem cells for the treatment of osteoarthritis of the knee: A 2-year follow-up study. The American Journal of Sports Medicine. 2017;**45**(12):2774-2783

[102] Garay-Mendoza D, Villarreal-Martínez L, Garza-Bedolla A, Pérez-Garza DM, Acosta-Olivo C, Vilchez-Cavazos F, et al. The effect of intra-articular injection of autologous bone marrow stem cells on pain and knee function in patients with osteoarthritis. International Journal of Rheumatic Diseases. 2018;**21**(1):140-147

[103] Kuah D, Sivell S, Longworth T, James K, Guermazi A, Cicuttini F, et al. Safety, tolerability and efficacy of intra-articular Progenza in knee osteoarthritis: A randomized double-blind placebo-controlled single ascending dose study. Journal of Translational Medicine. 2018;**16**(1):49

[104] Emadedin M, Labibzadeh N, Liastani MG, Karimi A, Jaroughi N, Bolurieh T, et al. Intra-articular implantation of autologous bone marrow-derived mesenchymal stromal cells to treat knee osteoarthritis: A randomized, triple-blind, placebo-controlled phase 1/2 clinical trial. Cytotherapy. 2018;**20**(10):1238-1246

[105] Khalifeh Soltani S, Forogh B, Ahmadbeigi N, Hadizadeh Kharazi H, Fallahzadeh K, Kashani L, et al. Safety and efficacy of allogenic placental mesenchymal stem cells for treating knee osteoarthritis: A pilot study. Cytotherapy. 2019;**21**(1):54-63

[106] Jevotovsky DS, Alfonso AR, Einhorn TA, Chiu ES. Osteoarthritis and stem cell therapy in humans: A systematic review. Osteoarthritis and Cartilage. 2018;**26**(6):711-729

[107] Marx RG, Wilson SM, Swiontkowski MF. Updating the assignment of levels of evidence. The

Journal of Bone and Joint Surgery. American Volume. 2015;**97**(1):1-2

[108] Vinardell T, Sheehy EJ, Buckley CT, Kelly DJ. A comparison of the functionality and in vivo phenotypic stability of cartilaginous tissues engineered from different stem cell sources. Tissue Engineering. Part A. 2012;**18**(11-12):1161-1170

[109] De Bari C, Dell'Accio F, Tylzanowski P, Luyten FP. Multipotent mesenchymal stem cells from adult human synovial membrane. Arthritis and Rheumatism. 2001;**44**(8):1928-1942

[110] Choi WH, Kim HR, Lee SJ, Jeong N, Park SR, Choi BH, et al. Fetal cartilage-derived cells have stem cell properties and are a highly potent cell source for cartilage regeneration. Cell Transplantation. 2016;**25**(3):449-461

[111] Lohan P, Treacy O, Griffin MD, Ritter T, Ryan AE. Anti-donor immune responses elicited by allogeneic mesenchymal stem cells and their extracellular vesicles: Are we still learning? Frontiers in Immunology. 2017;**8**:1626

[112] https://www.fda.gov/regulatory-information/search-fda-guidance-documents/osteoarthritis-structural-endpoints-development-drugs

[113] Yubo M, Yanyan L, Li L, Tao S, Bo L, Lin C. Clinical efficacy and safety of mesenchymal stem cell transplantation for osteoarthritis treatment: A meta-analysis. PLoS One. 2017;**12**(4):e0175449. Hills RK, editor

[114] Gold GE, Chen CA, Koo S, Hargreaves BA, Bangerter NK. Recent advances in MRI of articular cartilage. American Journal of Roentgenology. 2009;**193**(3):628-638

[115] Gentile P, Chiono V, Carmagnola I, Hatton P. An overview of poly(lactic-co-glycolic) acid (PLGA)-based biomaterials for bone tissue engineering. International Journal of Molecular Sciences [Internet]. 2014;**15**(3):3640-3659. Available from: http://www.mdpi.com/1422-0067/15/3/3640

[116] Tseng WJ, Huang S-W, Fang C-H, Hsu L-T, Chen C-Y, Shen H-H, et al. Treatment of osteoarthritis with collagen-based scaffold: A porcine animal model with xenograft mesenchymal stem cells. Histology and Histopathology [Internet]. 2018;**33**(12):1271-1286. Available from: http://www.ncbi.nlm.nih.gov/pubmed/29905361

[117] Wang B, Liu W, Xing D, Li R, Lv C, Li Y, et al. Injectable nanohydroxyapatite-chitosan-gelatin micro-scaffolds induce regeneration of knee subchondral bone lesions. Scientific Reports [Internet]. 2017;**7**(1):16709. Available from: http://www.ncbi.nlm.nih.gov/pubmed/29196647

[118] Deng C, Zhu H, Li J, Feng C, Yao Q, Wang L, et al. Bioactive scaffolds for regeneration of cartilage and subchondral bone interface. Theranostics [Internet]. 2018;**8**(7):1940-1955. Available from: http://www.ncbi.nlm.nih.gov/pubmed/29556366

[119] Mohammadi F, Mohammadi Samani S, Tanideh N, Ahmadi F. Hybrid scaffolds of hyaluronic acid and collagen loaded with prednisolone: An interesting system for osteoarthritis. Advanced Pharmaceutical Bulletin [Internet]. 2018;**8**(1):11-19. Availablee from: http://www.ncbi.nlm.nih.gov/pubmed/29670834

[120] de Barros CN, Miluzzi Yamada AL, Junior RSF, Barraviera B, Hussni CA, de Souza JB, et al. A new heterologous fibrin sealant as a scaffold to cartilage repair—Experimental study and

preliminary results. Experimental Biology and Medicine (Maywood, N.J.) [Internet]. 2016;**241**(13):1410-1415. Available from: http://www.ncbi.nlm.nih.gov/pubmed/26264444

[121] https://clinicaltrials.gov/ct2/show/NCT03800810

[122] Bansal H, Comella K, Leon J, Verma P, Agrawal D, Koka P, et al. Intra-articular injection in the knee of adipose derived stromal cells (stromal vascular fraction) and platelet rich plasma for osteoarthritis. Journal of Translational Medicine. 2017;**15**(1):141

Chapter 5

GMSC: Updates of Advances on Its Therapy in Immunological Diseases

Yuluan Hou and Song Guo Zheng

Abstract

Mesenchymal stem cells (MSCs) derived from various tissues are multi-potency of self-renewal and differentiation into multi-lineages, including chondrocytes, adipocytes and osteoblasts *in vitro* and *in vivo*. In addition, these cells also display potent immune regulatory roles that benefit the treatment of inflammatory and autoimmune diseases. We and others have previously identified that human gingival-derived mesenchymal stem cells (GMSCs) not only share similar biological features, but also display some potential advantages compared to other MSC populations. In the chapter, we have discussed the discovery, phenotypic and functional characteristics, as well as updated the advances of these cell therapies in immunological diseases.

Keywords: mesenchymal stem cells, gingival-derived mesenchymal stem cells, inflammatory and autoimmune diseases, immunomodulatory, cell therapy

1. Introduction

Mesenchymal stem cells (MSCs) are pluripotent stem cells derived from mesoderm with features of self-renewal and multi-lineage differentiation. These populations include MSCs that are primarily isolated from bone marrow (BMSCs) [1], fat [2], umbilical cord [3], dental pulp [4] and others [5, 6], particularly in most of adult tissues. Investigators have reported the successful differentiation of MSCs into mesenchymal-like cells, such as osteoblasts, chondrocytes, adipocytes [1–6], neural crest stem-like cells [7] and synoviocytes [8], and manifested that MSCs maintain immune homeostasis and prevent autoimmunity involving in the repair of impaired tissues and immunoregulation of autoimmune and inflammatory diseases [9, 10]. However, the occurrence and development of some autoimmune diseases are related to MSCs abnormality [11]. In addition, application of cell therapy using MSCs has weaknesses, like limited large-scale expansion *in vitro* and *vivo* [12], immunological rejection of allogeneic transplant and potential risk on tumorigenesis [13]. The availability of human gingival mesenchymal stem cells (GMSCs) together with their potent capacity of self-renewal, proliferation, multi-directional differentiation, inflammatory modulation and less tumorigenesis makes it as an ideal subtype of MSCs.

2. Discovery, development and biological characteristic of GMSCs

Gingiva is a unique oral tissue attached to the alveolar bone of tooth sockets, recognized as a biological mucosal barrier and a distinct component of oral mucosal immunity [14]. Wound healing within gingiva is characterized by rapid and fetal-like scarless healing, contrary to the common scar formation in skin [15]. Gingival tissue is easily assessable and gingival cells can be easily isolated and expanded from patients or healthy donors. Gingival fibroblast-like cells, including fibrocytes, myofibroblasts, pericytes and mesenchymal stem cells, a heterogeneous group of cells with distinct properties and functions, were named gingival fibroblasts before 2009, playing key roles in tissue development, maintenance and repair, as well as contributing to various pathologies [16]. Zhang et al. was the first to isolate and characterize a new population of precursor cells from human gingival tissues, termed GMSCs, which exhibit three unique stem cell-like properties as MSCs derived from bone marrow and other postnatal tissues [17]. Based on the minimal characterization criteria for human MSCs of the International Society for Cellular Therapy [18], the population of GMSCs shows: (1) *in vitro* proliferation as plastic-adherent cells with fibroblast-like morphology, colony-forming ability, (2) multipotent differentiation into different cell lineages, (3) positive expression of MSCs surface markers and stem cell-specific genes, negative of hemopoietic stem cell ones. Moreover, some studies reported that GMSCs manifested a higher expansion and telomerase activity and kept the features of MSCs, morphology and normal karyotype stable during cell expansion [19]. Recently, Gugliandolo et al. reported that most oncogenes of GMSCs at higher passengers were turned off, suggesting that long-term cultured GMSCs may be safer in the clinical setting [20].

Advantages	GMSCs	Other MSCs	References
Source	Gingiva a. Easily accessible and no invasive b. Wound healing rapid and scarless	Bone marrow and other adult tissues a. Invasive and rare from bone marrow and tendon b. Wound healing with scar in skin	[15]
Safty	Oncogenes at higher passengers turned off And no tumorigenic	Potential risk on tumorigenesis	[20]
Basical characters	Homogenous Proliferatation faster and morphology stable Normal karyotype and telomerase acticity in long-term cultures	Heterogenous Variable morphology and replicative senescence in vitro serial propagation Loose MSC characteristic at higher passages	[19]
Regulation of immune	Equal or even better immunoregulation in immunological diseases	Efficacy of the administration of MSCs for treatment of immune-related diseases has been confirmed in vivo studies.	[9], [10], [31], [53], [59], [63],
Positive factors	3D spheroid culture Hypoxic stimulation	No show in this chapter	[69],[74]

Table 1.
Advantages of GMSCs in the treatment of immunological diseases compared to other MSCs according to the updated studies.

Table 1 summarizes the advantages of GMSCs obtained from current studies in the treatment of immunological diseases.

3. Roles and mechanisms of GMSCs in treating immunological diseases

In response to the current challenges in the field of medicine on the efficacy and serious adverse effects of the current treatments, researchers are investigating the alternative therapies. In this regard, the use of MSCs represents a great promise for the treatment of a variety of immune-related diseases due to their potent properties of immunomodulatory ability [21, 22]. BMSCs and umbilical-cord MSCs (UC-MSCs) are widely studied because of their low immunogenicity [12, 23]. GMSCs, as a new subtype of MSCs, share strong abilities of immune regulation as MSCs from other tissues [24, 25]. As to update these advances of researches in GMSCs, we summarize the current recognition on the curative effects and mechanisms on immune and inflammation-related diseases (**Table 2**).

3.1 Immunoregulatory properties of GMSCs on autoimmune diseases

Autoimmune diseases are caused by defects in immune tolerance, resulting in that the body immune system failing to identify cells from their own or the foreign accompanied by the cellular or tissue damage [26]. Autoimmune diseases are categorized into systemic or organ-specific types according to the extent of tissue involvements [27]. The pathogenesis of autoimmune diseases is still not well-understood as multifactorial factors may be involved in at least both genetic and environmental factors [28, 29]. Although the conventional and biological therapies can somehow ameliorate clinical symptoms and decrease the morbidity and mortality, the limited efficiency, bone marrow toxicity and other side effects including infection and tumor are problematic [30]. Therefore it is desirable to find new strategies that can cure autoimmune diseases with minimal side effects. The MSCs therapy has been demonstrated to be likely as a new alternative approach. As a subset of MSCs, current studies show that GMSCs have even strong and better immunoregulatory effects than MSCs derived from other sources, through cell-cell contacting or secreting molecules to modulate both innate and adaptive responses [31].

Type 1 Diabetes Mellitus (T1DM) is an autoimmune disorder resulted from T cell-mediated destruction of pancreatic β-cells [32]. Hu et al. reported a clinical trial that implantation of Wharton's jelly-derived mesenchymal stem cells (WJ-MSCs) from the umbilical cord for newly-onset T1DM restored the function of islet β-cells in a longer time by improving the level of HbA1c and C peptide without acute or chronic side effects, suggesting that the implantation of WJ-MSCs for the treatment of newly-onset T1DM is safe and effective [33]. Researches in experimental models of mice manifested that MSCs inhibited the expansion of Th1, Th17 cells and stimulated the proliferation of CD4+CD25+Foxp3+ T regulatory (Treg) cells by reducing the levels of pro-inflammatory cytokines such as TNF-α, IFN-γ, CCL2, IL-1β, IL-2 and IL-17 but increasing the expression of immunoregulatory cytokines such as IL-4, IL-10 and IL-13 [9]. Zhang et al. infused GMSCs to determine the therapeutic effect on T1DM model, just as other MSCs, showed that GMSCs administration, homing to pancreas lymph nodes and pancreas, could delay diabetes onset, ameliorate pathology in pancreas by regulating down IL-17 and IFN-γ of CD4+ and CD8+ T cell and induce the generation of induced regulatory T (iTreg) *in vivo* which may be regulated through CD39/CD73 signal pathway [34]. Treg cells are a crucial immune suppressor that maintains the immune tolerance and prevents

Type of diseases		Model	Laboratory effects	Proposed mechanisms	References
Autoimmune diseases	T1DM	Mice	↓Diabetes onset ↓Pathology scores	↓IL-17 in spleen and lymph node ↓IFN-γ in spleen and lymph node ↑CD4+Foxp3GFP+Tregs ↑Islet β-cells	[34]
	CIA	Mice	↓Severity of arthritis ↓histopathology scores	↓IL-17 in lymph node ↓TGF-α in lymph nodes ↓IFN-γ in lymph nodes ↓IL-2 in lymph nodes ↑CD4+CD39+FoxP3+ Treg ↑CD39/CD73 signal ↑T-cell apoptosis	[38] [41]
Allo-GVHD		Mice	↓weight loss ↓inflammation degrees ↓pathological changes	↓T cell proliferation ↓IFN-γ, IL-4, IL-17 ↑CD39/CD73 signal ↑IDO signal	[45]
	Inflammatory bowel	Mice	↓diarrhea and weight loss ↓disease activity ↑restore injured mucosal tissues	↓CD4+T infiltration ↑IDO, IL-10 ↑Fas/FasL signal ↑Treg infiltration ↑T-cell apoptosis	[17], [49]
Inflammatory diseases	Atherosclerosis	Mice	↓plaque size and lipid deposition	↓recruitment of inflammatory macrophages ↓Inflammatory Ly-6Chi+monocytes ↓Macrophage Foam Cell Formation ↓M1 Macrophages ↓IFN-γ, IL-4 in spleen ↑M2 Macrophages	[51]
	Periodontitis	Dog Pig	↑cementum-like tissue ↑bone ↑sharpey fibers	↑differentiation of osteoblasts, cementoblasts, PDL fibroblasts in vivo ↑histological attachment level ↑junctional epithelium length ↑connective tissue adhesion	[55], [56]
Allergic diseases	CHS	Mice	↑desensitize ↑suppress CHS	↓ dendritic cells, $CD8^+$ T, TH-17 and MCs in LN ↑PGE2–EP3 signaling	[60],[61]
Wound healing		Mice	↑regenerated epidermis, skin appendages and hair follicles	↓M1 macrophages ↑M2 macrophages ↑re-epithelialization, collagen deposition and angiogenesis ↑TGF-β, CTGF,Tenascin-C ↑IL-1RA	[39],[40] [64],[65],

Table 2.
List of studies in which the therapeutic potential of administration of GMSCs for the treatment of immunological diseases was obtained.

the autoimmune responses [35–37]. It was also observed in rheumatoid arthritis animal model where GMSCs promoted Treg cell development to control autoimmune arthritis [38]. Some investigators also identified that high exocytotic fusion by secreting exosomes and cytokines is an alternative mechanism of GMSCs to promote the wound healing in diabetic patients, one of the most challenging complications in clinical medicine [39, 40].

The utilization of MSCs has reduced both the severity of disease and histopathology scores in rheumatoid arthritis models [9]. In collagen-induced arthritis (CIA) models, Chen et al. demonstrated that the adoptive transfer of GMSCs significantly delayed the onset of CIA and decreased the severity scores [38]. Histological and quantitative analysis of ankle joints demonstrated a significant decrease in synovitis, pannus formation and destruction of bone and cartilage in treated mice by increasing iTreg cells frequency while reducing percentages of Th1 and Th17 cells and relevant pro-inflammatory cytokines IFN-γ, IL-17, and TNF-α. Interestingly, Th2-type IL-4, IL-5 and IL-13 were not affected [38]. They found that GMSCs exerted the immune suppression functions indirectly *via* adenosine through CD39/CD73 signaling [38]. Recently, Gu et al. further supported this finding that GMSC ameliorated CIA and revealed that GMSC mediated T-cell apoptosis and influenced the polarization of the Th cells *via* a FasL/Fas pathway, resulting in immune tolerance and ameliorating the severity of CIA in mice [41].

3.2 Immunoregulation of GMSCs on graft-versus-host disease

Allogenic graft-versus-host disease (allo-GVHD) is a severe complication of organ or bone marrow transplantation related to the activation of alloreactive T cells or autoreactive mechanisms [42–44]. In clinical trials and experimental models, the administration of MSCs from bone marrow, adipose tissue and others decreased the severity of the symptoms and increased the survival. Most studies reported that MSCs inhibit reactive T cells trafficking and their proliferation. In addition, MSCs also stimulate cells differentiation into immunomodulatory cells such regulatory dendritic cells, Treg, Breg cells and M2 macrophages [9]. There is only a research of GMSCs for the treatment of allo-GVHD in mice model. Huang et al. revealed that GMSCs displayed the superior effect to BMSCs on suppressing xeno-GVHD according to the weight loss and inflammatory pathology in liver, lung, and intestine [45]. The underlying mechanism may be that GMSCs inhibited lymphocytes proliferation through CD39/CD73/adenosine and/or IDO signals without influencing CD4+Foxp3+ Treg cells [45].

3.3 Therapeutic progression of GMSCs in other inflammatory diseases

Inflammatory bowel disease characterized by dysfunction of the innate and adaptive immunity is a group of inflammatory conditions of the colon and small intestine [46–48]. The existing studies demonstrated administration of MSCs inhibited the proliferation and infiltration of inflammatory cells, for instance, significant inhibition in the expansion of Th1 and Th17 cells and opposite effect in the clonal expansion of Treg, by two main ways: direct cell-cell contact and the release of soluble factors [9]. In line with other tissues-derived MSCs, systemic infusion of GMSCs protected mice from colitis related tissue injuries and reduced the overall disease severity. Zhang et al. confirmed that GMSCs suppressed CD4+ T lymphocyte and promoted regulatory T cells infiltration to the colonic sites, which was accelerated by IFN-β-induced IDO and IL-10 [17]. While Yang et al. exhibited other actions of GMSCs in colitis that hydrogen sulfide upregulated the expression of

Fas/FasL in GMSCs coupling-induced T cells migration and T-cell apoptosis to maintain immunomodulation of GMSCs *in vivo* and *in vitro* [49].

Atherosclerosis is the major cause of cardiovascular diseases. Current evidences indicate that inflammation is involved in the pathogenesis of atherosclerosis and monocytes/macrophages are the major inflammatory cells [50]. Zhang et al. firstly indicated that GMSCs decreased inflammatory level, plaque size and lipid deposition in mice model *in vivo*, partly by inhibiting macrophage foam cell formation, suppressing the activation of M1 macrophages and promoting their development into the M2 phenotype *via* IDO and CD73 signals [51].

Periodontitis is a widespread bacterially induced immune-inflammatory disorder of the periodontium, featured with a progressive destruction of the tooth-supporting structures [52]. The milieu of bacterial biofilms challenges and activates host innate and adoptive immune systems to produce pro-inflammatory cytokines and chemokines for inflammatory cells recruitment, striking the balance of osteoblast and osteoclast [53]. MSCs from bone marrow, adipose, dental pulp and periodontal ligament have been testified to, *in vivo*, newly form periodontal bones, collagen fibers, periodontal ligament-like tissue and cementoid tissue indicating periodontium regeneration [54]. GMSCs also were proved to generate new cementum-like tissue, bone and sharpey fibers in dog and pig model of periodontitis [55, 56].

3.4 The contribution of GMSCs to contact hypersensitivity

Murine contact hypersensitivity (CHS) as a model similar to human allergic contact is caused by delayed-type hypersensitivity responses to antigens that come into contact with the skin [57]. The pathological process consists of sensitization phase, the elicitation or challenge phase, and resolution/regulation phase [58]. In this process, allergen-specific effector T cells and various types of innate immune cells are involved [59]. In 2011, Su et al. investigated the immunoregulatory role of GMSCs and for first time found that *i.v.* injection of GMSC significantly attenuated the CHS appearance at different phases of CHS, and showed that GMSCs-derived PGE2 played a crucial role in their inhibitory effect on dendritic cells and mast cells [60]. Li et al. further testified that PGE2–EP3 signaling played an important role in the immunomodulatory functions of GMSCs in murine CHS [61].

3.5 Wound healing

Cutaneous wound healing involves in three phases: inflammation, tissue formation, and remodeling [62]. Studies have demonstrated that systemically injected MSCs can home to injury sites accelerating wound repair [63]. Because of the rapid and fetal-like healing of gingival trauma, researchers have focused on the effect and mechanism of GMSCs. Experiments *in vitro* suggested that GMSCs were capable of switching macrophages from classical activation or proinflammatory M1 phenotype to an anti-inflammatory profile of M2 macrophages by soluble factors such IL-6, COX-2 and GM-CSF [64]. *In vivo* mice model suggested that enhancement of wound healing by systemic infusion of GMSCs related to enhanced re-epithelialization, collagen deposition and angiogenesis [64]. Compared with BMSCs, Linard et al. reported that gingival fibroblasts (GFs) intradermally injected in irradiated skin induced earlier development of thick, fully regenerated epidermis, skin appendages and hair follicles [65]. GFs also modified expression of ECM-related gene, ECM components (tenascin-C and a-smooth muscle actin) and wound healing-related factors, like TGF-β1 and CTGF [65]. While the influence to macrophage recruitment and differentiation of GFs was in accordance with Zhang et al., other studies presented that GMSC-derived exosomes accelerated wound healing in a diabetic rat skin defect model [39, 40]. Kou et al. indicated that TNF-α-Fas/Fap-1 *via* the NF-κB pathway enhancing IL-1RA release in GMSCs participated in healing progress [40].

4. Analysis of factors influencing the function of GMSC

Only well preserved the comprehensive and stable features, GMSCs can be an alternative cell therapy to autoimmune and inflammation-related diseases. Many factors can disturb the functions of GMSCs. Su et al. reported disturbed oral microbiome weakened the wound healing of GMSCs through miR-21/Sp1/telomerase reverse transcriptase pathway [66]. The physical condition of donors is a key factor to properties of GMSCs. Assem et al. revealed that GMSCs exhibited a greater proliferation rate and higher surviving in normal individuals than the diabetic patients [67]. Moreover, GMSCs exosome from diabetic mice showed reduced IL-1RA and decreased Fas expression when compared to WT GMSCs [40]. Different culture techniques of GMSCs have a profound effect on their biological functions. Subbarayan et al. showed that GMSCs derived spheroids enhanced abilities of viability, pluripotency and multi-lineages and maintained the properties of stem cells convincingly than conventional culture methods [68]. Zhang et al. have confirmed that spheroid-derived GMSCs possessed better therapeutic efficacy than their adherent counterpart [69]. The spheroid-derived GMSCs also had a greater homing ability to mucositis sites and underwent a higher mesenchymal-epithelial transformation compared to conventional culture GMSC in murine model of chemotherapy-induced oral mucositis [69]. Although normal and inflammatory GMSCs similarly expressed mesenchymal stem cells markers and proliferation ability, inflammatory microenvironments indeed reduced differentiation potentials of GMSCs [70]. Zhang et al. demonstrated that initial inflammatory stimuli of IL-1 and TNF-α appeared essential for GMSCs proliferation and tissue regeneration, while with inflammatory persistence, this effect turned to osteogenesis followed by a short-term stimulatory [71]. However, Apatzidou et al. demonstrated that GMSCs from periodontal granulation tissue possessed similar immunophenotype and regeneration feature to those in healthy periodontal tissue [72]. Many studies have also demonstrated that other elements also affect the biological characteristics of GMSCs, for instance, Lee et al. reported that dexamethasone accelerated the aging of GMSCs through downregulating SIRT1 and IL6 and upregulating EDN1 genes *via* the AGE/RAGE pathway [73]. In addition, hypoxic enhanced the suppressive effects of GMSCs on peripheral blood mononuclear cells and inhibited the local inflammation of injured skin by suppressing the inflammatory cells, accompanying with reduction of TNF-α and increase of IL-10 [74].

5. Conclusion

The existing studies have documented that GMSCs have self-renewal, multi-lineage differentiation potential, and immunomodulatory properties. These properties make GMSCs an alternative cell-based therapy of autoimmune and inflammation-related diseases. Plenty of internal and external factors may affect their functions of renewal, regeneration and immunoregulation. Moreover, the specific mechanisms and clinical efficacies are indistinct. Future studies and clinical trials should be implemented to elaborate mechanisms and therapeutic effects of immunomodulatory properties in detail on various inflammatory and immunological diseases.

Author details

Yuluan Hou[1] and Song Guo Zheng[2]*

1 Department of Clinical Immunology, Third Affiliated Hospital, Sun Yat-sen University, Guangzhou, China

2 Department of Internal Medicine, Ohio State University College of Medicine, Columbus, OH, USA

*Address all correspondence to: songguo.zheng@osumc.edu

References

[1] Caplan AI. Mesenchymal stem cells. Journal of Orthopaedic Research. 1991; **9**(5):641-650. DOI: 10.1002/jor.1100090504

[2] Zuk PA, Zhu M, Mizuno H. Multilineage cells from human adipose tissue: implications for cell-based therapies. Tissue Engineering. 2001; 7(2):211-228. DOI: 10.1089/107632701300062859

[3] Wang H-S, Hung S-C, Peng S-T, et al. Mesenchymal stem cells in the Wharton's jelly of the human umbilical cord. Stem Cells. 2004;**22**(7):1330-1337. DOI: 10.1634/stemcells.2004-0013

[4] Gronthos S, Brahim J, Li W, et al. Stem cell properties of human dental pulp stem cells. Journal of Dental Research. 2002;**81**(8):531-535. DOI: 10.1177/154405910208100806

[5] In't Anker PS, Scherjon SA, Kleijburg-van der Keur C, et al. Isolation of mesenchymal stem cells of fetal or maternal origin from human placenta. Stem Cells. 2004;**22**(7):1338-1345. DOI: 10.1634/stemcells.2004-0058

[6] Salingcarnboriboon R, Yoshitake H, Tsuji K, et al. Establishment of tendon-derived cell lines exhibiting pluripotent mesenchymal stem cell-like property. Experimental Cell Research. 2003; **287**(2):289-300

[7] Zhang Q, Nguyen PD, Shi S, et al. Neural crest stem-like cells non-genetically induced from human gingiva-derived mesenchymal stem cells promote facial nerve regeneration in rats. Molecular Neurobiology. 2018; **55**(8):6965-6983. DOI: 10.1007/s12035-018-0913-3

[8] Ferre FC, Larjava H, Loison-Robert LS, et al. Formation of cartilage and synovial tissue by human gingival stem cells. Stem Cells and Development. 2014;**23**(23):2895-2907. DOI: 10.1089/scd.2013.0547

[9] Leyendecker A Jr, Pinheiro CCG, Amano MT, et al. The use of human mesenchymal stem cells as therapeutic agents for the in vivo treatment of immune-related diseases: A systematic review. Frontiers in Immunology. 2018; **9**:2056. DOI: 10.3389/fimmu.2018.02056

[10] Ng J, Hynes K, White G, et al. Immunomodulatory properties of induced pluripotent stem cell-derived mesenchymal cells. Journal of Cellular Biochemistry. 2016;**117**(12):2844-2853. DOI: 10.1002/jcb.25596

[11] Sun LY, Zhang HY, Feng XB, et al. Abnormality of bone marrow-derived mesenchymal stem cells in patients with systemic lupus erythematosus. Lupus. 2007;**16**(2):121-128. DOI: 10.1177/0961203306075793

[12] Xu J, Wang D, Liu D, et al. Allogeneic mesenchymal stem cell treatment alleviates experimental and clinical Sjogren syndrome. Blood. 2012; **120**(15):3142-3151. DOI: 10.1182/blood-2011-11-391144

[13] Wang Y, Huso DL, Harrington J, et al. Outgrowth of a transformed cell population derived from normal human BM mesenchymal stem cell culture. Cytotherapy. 2005;7(6):509-519. DOI: 10.1080/14653240500363216

[14] Palmer R, Lubbock M. The soft connective tissues of the gingiva and periodontal ligament: Are they unique. Oral Diseases. 1995;**1**(4):230-237

[15] Mak K, Manji A, Gallant-Behm C, et al. Scarless healing of oral mucosa is characterized by faster resolution of inflammation and control of myofibroblast action compared to skin wounds in the red Duroc pig model.

Journal of Dermatological Science. 2009;**56**(3):168-180. DOI: 10.1016/j.jdermsci.2009.09.005

[16] Hakkinen L, Larjava H, Fournier BP. Distinct phenotype and therapeutic potential of gingival fibroblasts. Cytotherapy. 2014;**16**(9):1171-1186. DOI: 10.1016/j.jcyt.2014.04.004

[17] Zhang Q, Shi S, Liu Y, et al. Mesenchymal stem cells derived from human gingiva are capable of immunomodulatory functions and ameliorate inflammation-related tissue destruction in experimental colitis. Journal of Immunology. 2009;**183**(12):7787-7798. DOI: 10.4049/jimmunol.0902318

[18] Dominici M, Le Blanc K, Mueller I, et al. Minimal criteria for defining multipotent mesenchymal stromal cells. The International Society for Cellular Therapy position statement. Cytotherapy. 2006;**8**(4):315-317. DOI: 10.1080/14653240600855905

[19] Tomar GB, Srivastava RK, Gupta N, et al. Human gingiva-derived mesenchymal stem cells are superior to bone marrow-derived mesenchymal stem cells for cell therapy in regenerative medicine. Biochemical and Biophysical Research Communications. 2010;**393**(3):377-383. DOI: 10.1016/j.bbrc.2010.01.126

[20] Gugliandolo A, Rajan TS, Scionti D, et al. Reprogramming of oncogene expression in gingival mesenchymal stem cells following long-term culture in vitro. Cellular Reprogramming. 2017;**19**(3):159-170. DOI: 10.1089/cell.2016.0056

[21] Tanaka Y. Human mesenchymal stem cells as a tool for joint repair in rheumatoid arthritis. Clinical and Experimental Rheumatology. 2015;**33**(Suppl. 92):S58-S62

[22] Hynes K, Bright R, Proudman S, et al. Immunomodulatory properties of mesenchymal stem cell in experimental arthritis in rat and mouse models: A systematic review. Seminars in Arthritis and Rheumatism. 2016;**46**(1):1-19. DOI: 10.1016/j.semarthrit.2016.02.008

[23] Tsai PC, Fu TW, Chen YM, et al. The therapeutic potential of human umbilical mesenchymal stem cells from Wharton's jelly in the treatment of rat liver fibrosis. Liver Transplantation. 2009;**15**(5):484-495. DOI: 10.1002/lt.21715

[24] Fournier BP, Larjava H, Hakkinen L. Gingiva as a source of stem cells with therapeutic potential. Stem Cells and Development. 2013;**22**(24):3157-3177. DOI: 10.1089/scd.2013.0015

[25] Liu J, Yu F, Sun Y, et al. Concise reviews: Characteristics and potential applications of human dental tissue-derived mesenchymal stem cells. Stem Cells. 2015;**33**(3):627-638. DOI: 10.1002/stem.1909

[26] Ludwig RJ, Vanhoorelbeke K, Leypoldt F, et al. Mechanisms of autoantibody-induced pathology. Frontiers in Immunology. 2017;**8**:603. DOI: 10.3389/fimmu.2017.00603

[27] Marrack P, Kappler J, Kotzin BL. Autoimmune disease: Why and where it occurs. Nature Medicine. 2001;**7**(8):899-905. DOI: 10.1038/90935

[28] Edwards M, Dai R, Ahmed SA. Our environment shapes us: The importance of environment and sex differences in regulation of autoantibody production. Frontiers in Immunology. 2018;**9**:478. DOI: 10.3389/fimmu.2018.00478

[29] Ruocco V, Ruocco E, Lo Schiavo A, et al. Pemphigus: Etiology, pathogenesis, and inducing or triggering factors: Facts and controversies. Clinics in Dermatology. 2013;**31**(4):374-381. DOI: 10.1016/j.clindermatol.2013.01.004

[30] Liu Y, Pan YF, Xue YQ, et al. uPAR promotes tumor-like biologic behaviors of fibroblast-like synoviocytes through PI3K/Akt signaling pathway in patients with rheumatoid arthritis. Cellular & Molecular Immunology. 2018;**15**(2): 171-181. DOI: 10.1038/cmi.2016.60

[31] Huang F, Liu ZM, Zheng SG. Updates on GMSCs treatment for autoimmune diseases. Current Stem Cell Research & Therapy. 2018;**13**(5): 345-349. DOI: 10.2174/1574888X13666180220141114

[32] American Diabetes Association. Diagnosis and classification of diabetes mellitus. Diabetes Care. 2009;**32** (Supp. 1):S62-S67. DOI: 10.2337/dc09-S062

[33] Hu J, Yu X, Wang Z, et al. Long term effects of the implantation of Wharton's jelly-derived mesenchymal stem cells from the umbilical cord for newly-onset typ. 1 diabetes mellitus. Endocrine Journal. 2013;**60**(3):347-357

[34] Zhang W, Zhou L, Dang J, et al. Human gingiva-derived mesenchymal stem cells ameliorate streptozoticin-induced T1DM in mice via suppression of T effector cells and up-regulating Treg subsets. Scientific Reports. 2017; 7(1):15249. DOI: 10.1038/s41598-017-14979-5

[35] Liu ZM, Wang KP, Ma J, et al. The role of all-trans retinoic acid in the biology of Foxp3+ regulatory T cells. Cellular & Molecular Immunology. 2015;**12**(5):553-557. DOI: 10.1038/cmi.2014.133

[36] Li B, Zheng SG. How regulatory T cells sense and adapt to inflammation. Cellular & Molecular Immunology. 2015;**12**(5):519-520. DOI: 10.1038/cmi.2015.65

[37] Lan Q, Fan H, Quesniaux V, et al. Induced Foxp3(+) regulatory T cells: A potential new weapon to treat autoimmune and inflammatory diseases? Journal of Molecular Cell Biology. 2012;**4**(1):22-28. DOI: 10.1093/jmcb/mjr039

[38] Chen M, Su W, Lin X, et al. Adoptive transfer of human gingiva-derived mesenchymal stem cells ameliorates collagen-induced arthritis via suppression of Th1 and Th17 cells and enhancement of regulatory T cell differentiation. Arthritis and Rheumatism. 2013;**65**(5):1181-1193. DOI: 10.1002/art.37894

[39] Shi Q, Qian Z, Liu D, et al. GMSC-derived exosomes combined with a chitosan/silk hydrogel sponge accelerates wound healing in a diabetic rat skin defect model. Frontiers in Physiology. 2017;**8**:904. DOI: 10.3389/fphys.2017.00904

[40] Kou X, Xu X, Chen C. The Fas_Fap-1_Cav-1 complex regulates IL-1RA secretion in mesenchymal stem cells to accelerate wound healing. Science Translational Medicine. 2018;**10**(432): eaai8524

[41] Gu Y, Shi S. Transplantation of gingiva-derived mesenchymal stem cells ameliorates collagen-induced arthritis. Arthritis Research & Therapy. 2016; **18**(1):262. DOI: 10.1186/s13075-016-1160-5

[42] Chakraverty R, Sykes M. The role of antigen-presenting cells in triggering graft-versus-host disease and graft-versus-leukemia. Blood. 2007;**110**(1): 9-17. DOI: 10.1182/blood-2006-12-022038

[43] Huang F, Cao FL, Zheng SG. Update of humanized animal disease models in studying graft-versus-host disease. Human Vaccines & Immunotherapeutics. 2018;**14**(11):2618-2623. DOI: 10.1080/21645515.2018.1512454

[44] Lu L, Lan Q, Li Z, et al. Critical role of all-trans retinoic acid in stabilizing human natural regulatory T cells under inflammatory conditions. Proceedings

of the National Academy of Sciences of the United States of America. 2014; **111**(33):E3432-E3440. DOI: 10.1073/pnas.1408780111

[45] Huang F, Chen M, Chen W, et al. Human gingiva-derived mesenchymal stem cells inhibit xeno-graft-versus-host disease via CD39-CD73-adenosine and IDO signals. Frontiers in Immunology. 2017;**8**:68. DOI: 10.3389/fimmu.2017.00068

[46] Xavier RJ, Podolsky DK. Unravelling the pathogenesis of inflammatory bowel disease. Nature. 2007;**448**(7152):427-434. DOI: 10.1038/nature06005

[47] Gao Y, Tang J, Chen W, et al. Inflammation negatively regulates FOXP3 and regulatory T-cell function via DBC1. Proceedings of the National Academy of Sciences of the United States of America. 2015;**112**(25):E3246-E3254. DOI: 10.1073/pnas.1421463112

[48] Ramalingam R, Larmonier CB, Thurston RD, et al. Dendritic cell-specific disruption of TGF-β receptor II leads to altered regulatory T cell phenotype and spontaneous multiorgan autoimmunity. Journal of Immunology. 2012;**189**(8):3878-3893. DOI: 10.4049/jimmunol.1201029

[49] Yang R, Yu T, Liu D, et al. Hydrogen sulfide promotes immunomodulation of gingiva-derived mesenchymal stem cells via the Fas/FasL coupling pathway. Stem Cell Research & Therapy. 2018;**9**(1):62. DOI: 10.1186/s13287-018-0804-6

[50] Libby P. Inflammation in atherosclerosis. Nature. 2002; **420**(6917):868-874. DOI: 10.1038/nature01323

[51] Zhang X, Huang F, Li W, et al. Human gingiva-derived mesenchymal stem cells modulate monocytes/macrophages and alleviate atherosclerosis. Frontiers in Immunology. 2018;**9**:878. DOI: 10.3389/fimmu.2018.00878

[52] Page RC, Offenbacher S, Schroeder HE, et al. Advances in the pathogenesis of periodontitis: Summary of developments, clinical implications and future directions. Periodontology 2000. 1997;**14**:216-248

[53] Hajishengallis G. Periodontitis: From microbial immune subversion to systemic inflammation. Nature Reviews Immunology. 2015;**15**(1):30-44. DOI: 10.1038/nri3785

[54] Racz GZ, Kadar K, Foldes A, et al. Immunomodulatory and potential therapeutic role of mesenchymal stem cells in periodontitis. Journal of Physiology and Pharmacology. 2014; **65**(3):327-339

[55] Fawzy El-Sayed KM, Paris S, Becker ST, et al. Periodontal regeneration employing gingival margin-derived stem/progenitor cells: An animal study. Journal of Clinical Periodontology. 2012; **39**(9):861-870. DOI: 10.1111/j.1600-051X.2012.01904.x

[56] Yu X, Ge S, Chen S, et al. Human gingiva-derived mesenchymal stromal cells contribute to periodontal regeneration in beagle dogs. Cells, Tissues, Organs. 2013;**198**(6):428-437. DOI: 10.1159/000360276

[57] Honda T, Egawa G, Grabbe S, et al. Update of immune events in the murine contact hypersensitivity model: Toward the understanding of allergic contact dermatitis. The Journal of Investigative Dermatology. 2013;**133**(2):303-315. DOI: 10.1038/jid.2012.284

[58] Vocanson M, Hennino A, Rozieres A, et al. Effector and regulatory mechanisms in allergic contact dermatitis. Allergy. 2009;**64**(12):1699-1714. DOI: 10.1111/j.1398-9995.2009.02082.x

[59] Silvestre MC, Sato MN, Reis V. Innate immunity and effector and regulatory mechanisms involved in allergic contact dermatitis. Anais Brasileiros de Dermatologia. 2018;**93**(2): 242-250. DOI: 10.1590/abd1806-4841.20186340

[60] Su WR, Zhang QZ, Shi SH, et al. Human gingiva-derived mesenchymal stromal cells attenuate contact hypersensitivity via prostaglandin E2-dependent mechanisms. Stem Cells. 2011;**29**(11):1849-1860. DOI: 10.1002/stem.738

[61] Li P, Zhao Y, Ge L. Therapeutic effects of human gingiva-derived mesenchymal stromal cells on murine contact hypersensitivity via prostaglandin E2-EP3 signaling. Stem Cell Research & Therapy. 2016;**7**(1): 103. DOI: 10.1186/s13287-016-0361-9

[62] Gurtner GC, Werner S, Barrandon Y, et al. Wound repair and regeneration. Nature. 2008;**453**(7193):314-321. DOI: 10.1038/nature07039

[63] Agay D, Scherthan H, Forcheron F, et al. Multipotent mesenchymal stem cell grafting to treat cutaneous radiation syndrome: Development of a new minipig model. Experimental Hematology. 2010;**38**(10):945-956. DOI: 10.1016/j.exphem.2010.06.008

[64] Zhang QZ, Su WR, Shi SH, et al. Human gingiva-derived mesenchymal stem cells elicit polarization of m2 macrophages and enhance cutaneous wound healing. Stem Cells. 2010; **28**(10):1856-1868. DOI: 10.1002/stem.503

[65] Linard C, Tissedre F, Busson E, et al. Therapeutic potential of gingival fibroblasts for cutaneous radiation syndrome: Comparison to bone marrow-mesenchymal stem cell grafts. Stem Cells and Development. 2015; **24**(10):1182-1193. DOI: 10.1089/scd.2014.0486

[66] Su Y, Chen C, Guo L, et al. Ecological balance of oral microbiota is required to maintain oral mesenchymal stem cell homeostasis. Stem Cells. 2018; **36**(4):551-561. DOI: 10.1002/stem.2762

[67] Assem M, Kamal S, Sabry D, et al. Preclinical assessment of the proliferation capacity of gingival and periodontal ligament stem cells from diabetic patients. Open access Macedonian journal of medical sciences. 2018;**6**(2):254-259. DOI: 10.3889/oamjms.2018.076

[68] Subbarayan R, Murugan Girija D, Ranga Rao S. Gingival spheroids possess multilineage differentiation potential. Journal of Cellular Physiology. 2018; **233**(3):1952-1958. DOI: 10.1002/jcp.25894

[69] Zhang Q, Nguyen AL, Shi S, et al. Three-dimensional spheroid culture of human gingiva-derived mesenchymal stem cells enhances mitigation of chemotherapy-induced oral mucositis. Stem Cells and Development. 2012; **21**(6):937-947. DOI: 10.1089/scd.2011.0252

[70] Li N, Liu N, Zhou J, et al. Inflammatory environment induces gingival tissue-specific mesenchymal stem cells to differentiate towards a pro-fibrotic phenotype. Biology of the Cell. 2013;**105**(6):261-275. DOI: 10.1111/boc.201200064

[71] Zhang F, Si M, Wang H, et al. IL-1/TNF-alpha inflammatory and anti-inflammatory synchronization affects gingival stem/progenitor cells' regenerative attributes. Stem Cells International. 2017;**2017**:1349481. DOI: 10.1155/2017/1349481

[72] Apatzidou DA, Nile C, Bakopoulou A, et al. Stem cell-like populations and immunoregulatory molecules in periodontal granulation tissue. Journal of Periodontal Research. 2018;**53**(4): 610-621. DOI: 10.1111/jre.12551

[73] Lee H, Kim M, Park YH, et al. Dexamethasone downregulates SIRT1 and IL6 and upregulates EDN1 genes in stem cells derived from gingivae via the AGE/RAGE pathway. Biotechnology Letters. 2018;**40**(3):509-519. DOI: 10.1007/s10529-017-2493-0

[74] Jiang CM, Liu J, Zhao JY, et al. Effects of hypoxia on the immunomodulatory properties of human gingiva-derived mesenchymal stem cells. Journal of Dental Research. 2015;**94**(1):69-77. DOI: 10.1177/0022034514557671

Section 3

Specific Roles and Properties of Mesenchymal Stem Cells

Chapter 6

The Angiogenic Paracrine Potential of Mesenchymal Stem Cells

Jafar Rezaie, Morteza Heidarzadeh, Mehdi Hassanpour, Hassan Amini, Elhameh Shokrollahi, Mahdi Ahmadi and Reza Rahbarghazi

Abstract

Tissue engineering and regenerative medicine are branches of biomedical sciences that facilitate the use of cells and biocompatible scaffolds in favor of tissue restoration. In this regard, restoration and maintenance of angiogenesis and blood supplementation could be an effective strategy for injured tissue removal, accelerating healing rate, and successful transplantation of cells and scaffolds into target sites. It has been elucidated that mesenchymal stem cells have the potency to promote angiogenesis via paracrine activity and trans-differentiation into the endothelial lineage. In this chapter, we highlighted the paracrine property of mesenchymal stem cells to modulate angiogenesis in the target tissues.

Keywords: mesenchymal stem cells, angiogenesis, paracrine activity, exosomes

1. Introduction

Angiogenesis, termed as neovascularization, is defined as de novo vascularization from the pre-existing vascular network and activated in response to numerous pathological and physiological stimuli, playing critical roles during development and tissue repair [1]. Recent advances in the field of stem cell research, notably MSCs, have opened new horizons to human medicine in the promotion of angiogenesis and restoration and salvage of ischemic tissues [2]. MSCs actively participate in angiogenesis via direct differentiation, cell contact interaction with endothelial lineage, and releasing pro-angiogenic factors via a paracrine manner [3]. Due to the low survival and differentiation rate of MSCs posttransplantation into ischemic microenvironment, it is proposed that the paracrine activity is the principal mechanism for the therapeutic outcome [4]. It has been well-established that stem cell-secreted growth factors are responsible for, at least in part, therapeutic effects. As a matter of fact, MSC-derived secretome is thought to be a suitable alternative therapeutic modality to MSCs posttransplantation. At present, the underlying mechanisms by which MSC secretomes contribute to tissue healing and angiogenesis are not fully addressed and many efforts are needed to fill knowledge gaps by experimental animal research and clinical trials prior to application to human medicine [5, 6]. Paracrine factors could increase the blood supplement of damaged tissues via the activation and recruitment of resident/circulating stem cells and progenitor cells [7, 8]. Several experiments detected the pro-angiogenic capacity of MSCs isolated from different sources [9, 10]. **Table 1** ELISA and liquid-chip assays of cytokine content of umbilical cord MSCs revealed several angiogenesis factors, including interleukin-8 (IL-8), insulin-like growth factor 1 (IGF-1), and vascular endothelial growth factor (VEGF) compared to mature cell types such as fibroblasts. These pro-angiogenic factors are able to form vascular networks and increase the migration of endothelial lineage in vitro [51]. In addition to the secretion of angiogenic factors by MSCs, it was revealed that various factors existing in secretome could activate the angiogenic behavior in endothelial cells (ECs). For instance, equine peripheral blood MSC angiocrine was found to stimulate endothelial functional behavior by the induction of VEGF-A signaling pathway via several factors such as endothelin-1, IL-8, platelet-derived growth factor-AA (PDGF-AA), and IGF-2 [52]. Due to the variety of factors released by MSCs such as VEGF, monocyte chemoattractant protein-1 (MCP-1), and IL-6, an increased angiogenesis rate was observed in the mouse model of hindlimb ischemia, and even the combination of VEGF, MCP-1, and IL-6 could be served as a commercial cocktail for the promotion of angiogenesis either in vivo or in vitro [53]. In addition to the existence of the pro-angiogenic factor in MSC secretome, some authorities, however, showed the anti-angiogenic properties of these cells (**Table 2**) [67]. In some circumstances, the dual effect of a distinct factor was proved related to angiogenesis status. For example, in VEGF-free condition, the attachment of angiopoietin-2 (Ang-2) to receptor tyrosine kinase (RTK), namely Tie-2, promotes vascular destabilization and regression by reduction of pericyte-EC interaction, while in normal condition Ang-2 could increase EC migration and tip cell formation required for neovascularization [68]. Commensurate with these comments, one could hypothesize that the dynamic balance of MSC secretome, cell source, purity, and preconditioning could predetermine the pro- and/or anti-angiogenic property of MSCs [67].

By modulating distinct signaling pathway/s inside the MSCs, cell bioactivity would be induced in favor of neovascularization. For instance, it was shown that the activation of sonic hedgehog (Shh) factor in Wharton's jelly-derived MSCs (WJ-MSCs) induced the production of pro-angiogenic factors such as angiogenin, angiopoietin-1, activin A, matrix metallopeptidase-9 (MMP-9), granulocyte-macrophage colony-stimulating factor, and urokinase-type plasminogen activator, indicating WJ-MSCs an ideal cell source for the induction of vascularization [69]. An experiment conducted by Matluobi et al. showed an enhanced vascular formation capacity of human MSCs after treatment with carvacrol evaluated by chicken chorioallantoic membrane angiogenesis assay. The carvacrol-treated MSCs tended to trans-differentiate into endothelial lineage by the expression of VWF and VE-cadherin [70]. MSCs have the ability to adapt themselves with environmental condition increasing regenerative potential in different conditions [71]. Maintaining the MSC cross talk with other cells is required for cell hemostasis, stemness feature, and regenerative potential in the distinct niche. For example, the normal bioactivity of Hox gene, *Abdominal-B*, seems to be essential in Drosophila cystic stem cells to obtain multipotentiality [72].

Regarding issues related to isolation protocols and stem cell proliferation rate, a careful selection is essential for high-throughput results. Vizoso et al. demonstrated large-scale secretome production and release of a vast array of bioactive factors in human uterine cervical stem cells with considerable advantages over MSCs from other tissues for research and clinical application [73].

The emergence of some conditions could change the trans-differentiation capacity of MSCs into distinct phenotypes. In the case of the vicious cycle of abnormal placental development in intrauterine growth restriction, placental mesenchymal stromal cells lose angiogenic potential while acquiring adipogenic capacity which is coincided with a metabolic shift from aerobic to anaerobic state [71]. It seems that

Factor	Amniotic fluid/ bone marrow MSCs	Function
Angiogenin	+/+	A pancreatic ribonuclease, known as ribonuclease 5, which induces vascularization [11]
Angiopoietin-1	+/+	Activates TEK/TIE2 receptor; promotes angiogenic processes, endothelial cell survival, migration, proliferation, and stabilization; and during embryogenesis has a role in heart development [12]
Angiopoietin-2	+/−	Binds to TEK/TIE2, in the presence of VEGF and Ang-2 and promotes neovascularization [13, 14]
Angiopoietin-4	+/+	Binds to TEK/TIE2, modulating ANGPT1 signaling, can induce tyrosine phosphorylation of TEK/TIE2, and promotes endothelial cell survival, migration, and angiogenesis [15]
Amphiregulin	+/−	An EGF-like ligand that binds to the EGFR, enhanced lymphangiogenesis, and stimulates the growth of normal epithelial cells [16]
Artemin	+/−	Binds for the GFR-alpha-3-RET and GFR-alpha-1-RET receptor and promotes angiogenesis [17]
Tissue factor	+/−	Stimulates PDGF receptor signaling pathway, angiogenesis, endothelial cell migration, chemotaxis and proliferation, and coagulation factor III/CD142; improves transcription of VEGF; and reduces transcription of the thrombospondins [18]
CXCL16	+/+	Encourages a chemotactic response, pro-angiogenic [19]
DPPIV	+/−	A membrane-bound oligopeptidase acting on and modulating the pro-angiogenic chemokine CXCL12 [20]
Epidermal growth factor	+/−	Encourages the growth of epithelial tissues, is anti-apoptotic, induces lymphangiogenesis, and improves MSC survival [21]
EG-VEGF	+/−	Also called Prokineticin 1. Binds to PROKR1 and PROKR2, pro-angiogenic [22]
Endothelin-1	+/+	Derived from the endothelium with vasoconstrictor and angiogenic effects, prolymphoangiogenic [23]
Endoglin	+/−	Also called CD 105. Modulates TGF-β1 and β3 responses, vascular development, and angiogenic effects [24]
FGF-7	+/+	Has positive effects on cell proliferation, migration and division, chemotaxis, and arteriogenesis [25]
Acidic FGF/FGF-1	+/−	Binds to for FGFR1 and integrins and induces angiogenesis [26]
Basic FGF/FGF-2	+/−	Ligand for FGFR1, FGFR2, FGFR3, and FGFR4, Vascular regeneration; role in cell migration and proliferation involved in angiogenesis, stimulates arteriogenesis [27]
FGF-4	+/−	Has positive effects in MSC proliferation, pro-angiogenic [28]
GDNF	+/−	Has positive effects in angiogenesis [29]
GM-CSF	+/−	Has positive effects in angiogenesis [30]
Heparanase	+/+	Has positive effects in angiogenesis [31]
Heparin binding-EGF	+/+	Has positive effects in angiogenesis [32]
Hepatocyte growth factor	+/−	Has positive effects in angiogenesis [33]
HIF-1α	+/+	Functions as a master transcriptional regulator of the adaptive response to hypoxia and influences cell metabolism, cell survival, and angiogenesis [34]

Factor	Amniotic fluid/ bone marrow MSCs	Function
IL-1β	+/−	Has positive effects in angiogenesis and lymphangiogenesis [35]
IL-6	+/−	A potent pro-angiogenic cytokine which stimulates endothelial cell and smooth muscle cell proliferation and migration and promotes neovascularization [36]
IL-8	+/−	Has a role of pro-angiogenic factor [37]
Leptin	+/−	Stimulates vessel formation [38]
MCP-1	+/−	CCL2. Induces stabilization of new vessels [39]
MIP-1α	+/−	CCL3. Induces vessel formation
MMP-8	+/−	Known as collagenase 2. Breaks collagen types I, II, and III and has positive effects on angiogenesis [40]
MMP-9	+/−	Called as gelatinase B. Breaks both collagens and gelatins and has positive effects on angiogenesis [41]
NRG1-β1	+/−	Promotes angiogenesis and arteriogenesis [42]
Pentraxin-3 (PTX3)	+/+	Has a role of a pro-angiogenic agent [43]
PD-ECGF	+/−	Stimulates angiogenesis [44]
PDGF-AA	+/+	Has positive effects on MSC proliferation and stimulates angiogenesis [45]
PDGF-AB/ PDGF-BB	+/−	Induces neovascularization and arteriogenesis [27]
Persefin	+/−	Induces angiogenesis [3]
PlGF	+/+	Has a role of a pro-angiogenic factor [46]
Prolactin	+/−	Has a role of a pro-angiogenic factor in intact form [47]
Sphingosine kinase 1	+/+	Promotes angiogenesis [48]
SDF-1α	-/+	An important chemotactic factor for progenitor cells. Stimulates stem cell migration, adhesion, and homing [3]
TGF-β1	+/−	Promotes angiogenesis at least in part via the secretion of the survival factors TGF-α and VEGF [3]
uPA	+/+	Promotes endothelial cell proliferation and migration and has positive effects in vascular network formation [49]
VEGF	+/+	Promotes angiogenesis [50]
VEGF-C	+/−	Promotes lymphangiogenesis [50]

Table 1.
Comparison of angiogenic paracrine factors secreted by MSCs from amniotic fluid and bone marrow.

external environmental influence could alter the therapeutic potency of MSCs by rendering epigenetic marks associated with cell differentiation capacity [74]. In support of this claim, Rezaie and co-workers found a decrease of angiogenic human MSC potential after exposure to diabetic sera. The diabetic MSCs showed a declined migration capacity by suppressing the transcription of MMP-2, MMP-9, and CXCR-4 and aborted the secretion of Ang-1, Ang-2, and VEGF [75]. The expression of CXC chemokine receptors such as CXCR-1, CXCR-2, and CXCR-4 was found to accelerate and direct MSC migration in response to the chemokine gradients. A blockade of CXCR chemokine such as CXCL6 had potential to abrogate cardiac stem cell migration and motility [76].

Factor	Amniotic fluid/ bone marrow (MSCs)	Function
Angiopoietin-2	+/−	Binds to TEK/TIE2 and induces endothelial cell apoptosis in the absence of VEGF [54, 55]
Angiostatin	+/−	Angiogenic inhibitor. Acts as an inhibitor of endothelial cell proliferation and migration [56]
Endostatin	+/−	Acts as inhibitor of endothelial cell proliferation and migration and angiogenesis and induces endothelial apoptosis [56]
TGF-β1	+/−	Angiogenic inhibitor [57]
Platelet factor 4 (PF4)	+/−	Angiogenic inhibitor [58]
Serpin B5	+/−	Maspin. A member of the serine protease inhibitor family and negative regulator of angiogenesis [59]
Serpin E1	+/+	Serine protease inhibitor; inhibition of angiogenesis; inhibitor of uPA; preserves the vascular integrity [60]
Serpin F1	+/+	Serine protease inhibitor, inhibition of angiogenesis [61]
TIMP-1	+/+	Angiogenesis inhibitor [62]
TIMP-4	+/+	Angiogenesis inhibitor [63]
Thrombospondin-1	+/+	Anti-angiogenic. Inhibits endothelial cell proliferation [64]
Thrombospondin-2	+/−	Anti-angiogenic. Inhibits endothelial cell migration and tubule formation [65]
Vasohibin	+/−	A negative feedback regulator of angiogenesis [66]

Table 2.
Comparison of anti-angiogenic paracrine factors secreted by MSCs from the amniotic fluid and bone marrow.

At present, the combination of cell and tissue engineering techniques increased the restoration potential of a distinct cell type after transplantation [77]. In most of these approaches, the maintenance of cell-to-cell interaction in 3D microenvironment could increase survival signaling pathway and organotypic plasticity of cells. For instance, it seems that cell encapsulation by the mixture of alginate-gelatin promotes angiocrine cues and vascular network formation [77]. The introduction of MSC-alginate microbeads to ischemic hindlimb mouse model promoted arterial collaterals after the occlusion of the femoral artery by the modulation of VEGF-A signaling pathway [78]. A side-by-side comparison of MSCs expanded in 2D, and alginate microbeads revealed enhanced angiogenic and chemotactic activity in cutaneous healing [79].

2. Modulation of angiogenesis by exosomes

Regarding paracrine activity, MSC exosomes transfer various bioactive molecules, microRNAs, and protein factors with the ability to modulate angiogenesis behavior in the target cells.

2.1 Exosomes biogenesis

Exosomes are a subtype of extracellular vesicles (EVs, 40–200 nm) found in bio-fluids and released from all cell types. They maintain cell-to-cell communication through shuttling diverse biomolecules [80–82]. The first intracellular step in exosome biogenesis involves the invagination of the membrane of the multivesicular body (MVB) to form membrane-bound vesicles in MVB lumens that are identified as intraluminal vesicles (ILVs) (**Figure 1**) [83, 84]. Various factors and signaling pathways have been considered in biogenesis, trafficking, and abscission of exosomes [85]. Of note, endosomal sorting complexes required for transport (ESCRT) machinery with four complexes, ESCRT-0, ESCRT-I, ESCRT-II, and ESCRT-III, participate in exosome formation and packing cargo incorporation with different accessory proteins (**Figure 1**) [81, 85, 86]. Noteworthy, the formation of MVBs in the absence of the ESCRT machinery is aborted. In this condition, oligodendroglial cell ceramide is a key molecule to induce inward budding of the limiting membrane of MVBs [83, 87]. After MVB formation, intracellular trafficking of vesicle systems was orchestrated by Rab-GTPase family proteins [81]. As shown in **Figure 1**, several Rab proteins specifically contribute to the transfer of vesicles in definitive pathways. Along with these factors, soluble NSF attachment protein receptor (SNARE) has been suggested to control the fusion of MVBs with the plasma membrane (**Figure 1**) [88]. At the intracellular level, three possible fates are considered to involve MVBs such as secretory, lysosomal, and back fusion pathways. Once secreted, exosomes can be received by neighboring cells by three possible

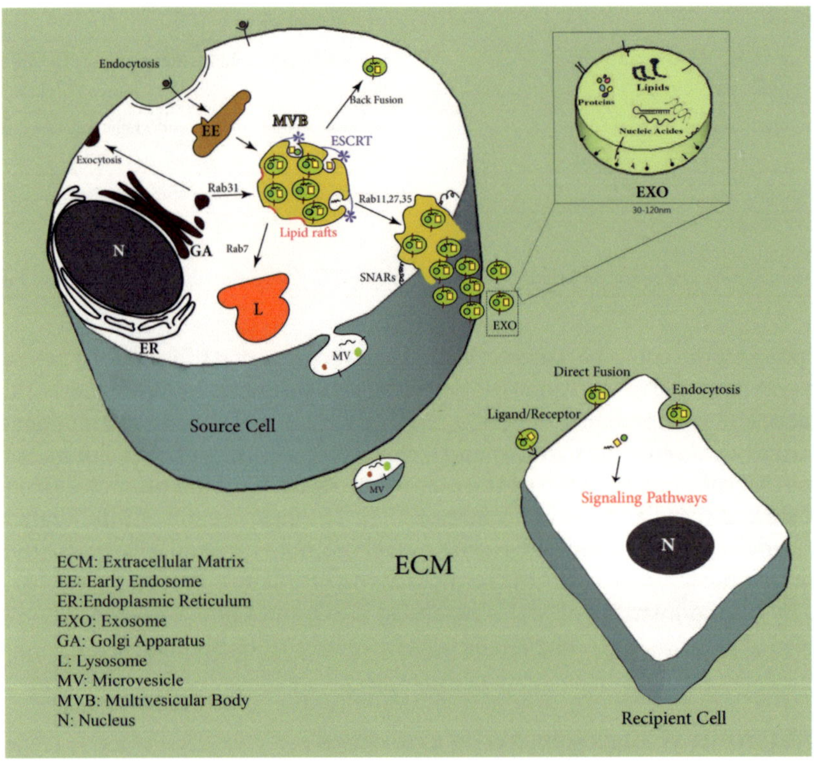

Figure 1.
Biogenesis, structure, and uptake of exosomes. Exosomes are producing during invagination process of MVB's membrane. ESCRT machinery and ESCRT-independent mechanisms (lipid rafts/tetraspanin) contribute to form exosomes and sort several molecules including proteins, miRNA, mRNA, DNA strands, and lipids into their lumen or limiting membrane of exosomes. Exosome cargoes are collected from materials received by endocytic pathway, Golgi apparatus, and cytoplasm. Rab-GTPase family proteins regulate intercellular trafficking and docking of MVBs. In the secretory pathway, MVBs actively fuse with the plasma membrane to release exosomes into the extracellular space. In alternative pathways, MVB could prefer binding to the lysosome or directly fuse back to the plasma membrane. Once secreted, exosomes enroll several mechanisms to arrive at the target cell: (I) enter through internalization process; (II) bind through receptor-ligand interactions, (III) direct fusion with the plasma membrane of the target cell. Exosomes are able to affect the biological processes of the target cells.

mechanisms: (i) internalization, (ii) direct fusion, and (iii) receptor-ligand interaction. Exosomal uptake results in triggering signaling pathways reprogramming fate, proliferation, survival, and morphology of recipient cells (**Figure 1**) [89, 90].

2.2 Pro- and anti-angiogenic capacity of exosomes

It was shown that a significant portion of MSC angio-activity drives from their potency to release exosomes that can affect the function of ECs, either by increasing the production of pro-angiogenic factors or decreasing the production of anti-angiogenic factors [91]. The fact that MSC exosomes promote angiogenesis, by delivering mediators such as miRNAs, protein factors to distinct cells, was confirmed in various in vivo experimental studies [89, 92, 93].

2.2.1 miRNAs

It seems that exosomal cargo such as cytokines and miRNAs could be easily transferred to recipient cells. Increasing evidence indicates that exosomal pro-angiogenic miRNAs (miRNA-125a, miRNA-30b, miRNA-30c, miRNA-424, miRNA-150, and let-7f) are important regulators of angiogenesis in the target sites [89, 94–96]. Data suggest that exosomal miR-150 is a key contributor to the pro-angiogenic activity of MSC exosomes following ischemic injuries [89, 96, 97]. In contrast, anti-angiogenic function on tumor cells was reported by a research group guided by Lee et al. They demonstrated the anti-angiogenic function of MSC exosomes on breast cancer cells governed by delivering miR-16 to suppress VEGF factor [91]. In a recent study conducted by Chen et al., they declared that exosome can be used as therapeutic transfer vesicles to carry miRNAs and genetic molecules to modulate VEGF content and control untamed angiogenesis in rheumatoid arthritis [98]. Based on the literature, the expression of VEGF, endothelial marker CD31, and matrix metalloproteinases-14 (MMP-14) activity is induced in patients with rheumatoid arthritis. The application of MSC-derived exosomes containing miRNA-150-5p (Exo-150) clearly decreased transcription of VEGF and MMP-14 in synovial fluid. Consistent with these changes, the pro-inflammatory response was blunted by decreasing IL-1β, transforming growth factor-β (TGF-β) and tumor necrosis factor-α (TNF-α) content in synovial fluid. This study has shown that MSC-derived Exo-150 can be used as bio-shuttle and magic bullet for inhibiting an exacerbated angiogenesis via the modulation of angiogenesis-related factors. However, some contradictory facts exist regarding the sole application of exosomes in the context of tumor cells.

2.2.2 Exosomal pro- and anti-angiogenic factors

MSCs can secret signal transducer and activator of transcription-3 (STAT3) mRNAs via exosomes that augment the transcription of hepatocyte growth factor (HGF), IL-6, and VEGF, promoting proliferation and migration of ECs [99]. In this context, MSC exosomes abundantly are enriched with VEGF factor that increases neovascularization through the Wnt4/β-catenin pathway in epithelial cells [100, 101]. The pro-angiogenic propriety of MSC exosomes has been previously shown in myocardial ischemia/reperfusion injury experiments following acute myocardial infarction [102–104]. In contrast, MSC exosomes may contain abundant anti-angiogenic factors that could regulate tumor angiogenesis rate. Lee et al. showed that exosomes from MSCs significantly downregulated the expression of VEGF in breast cancer cells, leading to the abortion of angiogenesis [91]. However, there are contradicting results. For example, human bone marrow MSC exosomes promoted VEGF synthesis in colonic and gastric carcinomas through the activation of extracellular signal-regulated kinase1/kinase2 (ERK1/ERK2) and p38 MAPK pathways [105]. Taken together, these issues show a fact that exosomes from various MSC types can mediate physiological and pathological angiogenesis and could be considered as a suitable bio-shuttle for establishing promising therapeutic approaches in an individual with cancers and ischemic pathologies. The feasibility of exosome uptake by recipient cells, make these cell products for introducing in clinical approaches. Xue and colleagues investigated the effects of cord blood and adipose-derived MSC exosomes on human EC angiogenesis capacity under hypoxic and normal conditions [106, 107]. They noted the potency of isolated exosomes in triggering angiogenesis rate especially under the hypoxic condition compared to exosome counterpart originated from normal milieu. Based on their data, the transcription level of genes related to angiogenesis such as angiopoietin-1 (Ang-1) and VEGF receptor-2 (also termed FLK-1) was induced significantly after exposure to exosomes collected from hypoxic MSCs rather than that of normal cells. Following the induction of Ang-1 and FLK-1, the status of some downstream effectors would be turned to an activated form. For instance, it was found that protein kinase A (PKA) is indirectly triggered after the activation of genes *Ang-1* and *VEGFR-2*. Along with changes, the transcription level of angiogenesis inhibitory gene like Vash1 is completely suppressed. The inhibitory angiogenesis potential of MSCs was investigated on cancer cells or progenitors residing inside tumor mass. Both anti-inflammatory and pro-angiogenesis property of MSC-derived exosomes were shown in cardiovascular disease [92, 97]. In addition to the promotion of cell surface receptors, exosomes could augment the synthesis of VEGF factor in targeted cells. Doeppner et al. also previously demonstrated that MSC-derived exosomes initiated healing processes after the onset of neurological diseases by increasing angiogenesis and blood supply which led to the neurological recovery and neurogenesis [108]. Other experiments added notion on the potency of exosomes to reduce neuroinflammation in traumatic brain injury [109]. However, some contradictory facts exist regarding the sole application of exosomes in the context of tumor cells. The superior stimulatory effect of MSC-derived exosomes on tumor angiogenesis was also addressed by different authors [110]. For example, Zhu et al. demonstrated the vasculogenic role of MSC exosomes after addition to human gastric carcinoma (SGC-7901) and colon cancer (SW480) cell lines [105, 111]. They found that the normal status of signaling effectors such as phosphorylated ERK1/ERK2, Bcl-2, and VEGF proteins; alpha-smooth muscle actin (α-SMA); CXCR-4; and mouse double minute 2 homolog (MDM2) mRNA was modulated in the favor of angiogenesis in a mouse cancer model. In addition to the direct fast action on recipient cells, it is reasonable to hypothesize that exosomes are able to dictate pro-/anti-angiogenesis pattern in distinct cells by provoking specific signaling pathways and effectors such as ERK1/ERK2 and p38 MAPK kinase routes.

The engagement of factors such as AKT, STAT3, Wnt/β-catenin, and ERK happens following cutaneous wound regeneration treated with MSC exosomes. Proteomic analysis revealed that the protein content of growth factors IL-6, stromal cell-derived factor-1α (SDF-1α), IGF-1α, STAT3, and HGF contributed to cell proliferation, migration, and angiogenesis, improving reepithelialization in wound sites [112]. The modulation of Wnt/β-catenin pathway targeting Wnt4 diminishes the number of cells with apoptotic changes with the levels of pro-angiogenic factors such as IL-6 and IL-8, granulocyte-colony-stimulating factor (G-CSF), PDGF-BB, MCP-1, and VEGF are increased. In response to treatment with exosomes, phosphorylation of glycogen synthase kinase 3β (GSK3β) as a main negative regulator of Wnt signaling pathway is initiated, resulting in the progression of a cell from phase G1 to S and cutaneous cell proliferation [113]. An enhanced angiogenesis rate with reduced cardiomyocyte apoptosis was reported following the administration of MSCs to infarct area. After being exposed to the ischemic/hypoxic condition, MSCs were programmed to secrete exosomes. Under these conditions, GATA-4 is induced which prevents cell apoptosis, reducing the infarct size. Meanwhile, the level of an anti-apoptotic agent such as miR-19a and miR-22 was increased in the target sites [114]. In another experiment conducted by Teng et al., it was shown that MSC-derived exosomes harboring miRNA-132 efficiently are delivered to human umbilical vein ECs (HUVECs). Therefore, it could be pointed out that MSCs could dictate prominent changes in the target cells. They also declared that endothelial Ras signaling pathway effectors are modulated by recipient cells after direct interaction of this miRNA with *RASA1* gene. Ras group genes have a basic role in controlling cell proliferation and differentiation [107]. Along with these statements, the bona fide effects of MSC exosomes need to be precisely addressed by a plethora of various experiments.

In the context of tumor niche, both anti- and pro-tumorigenic features was found after the treatment of cancer cells with MSCs exosomes. The migration and proliferation of tumor cells were tightly regulated by exosome factors by the modulation of PDGFR, C-Met, and EGFR signaling. Ex vivo modulation showed this fact that MSC exosomes could activate or phosphorylate intracellular kinase domain of relevant receptors, thereby triggering Akt, PKC/PKB, and MAP signaling pathways, leading to proliferation and migration of gastric tumor cells [115]. Exosomes released by human bone marrow MSCs augmented VEGF in colonic carcinoma and gastric carcinoma tumor cells through the activation of ERK1/ERK2 and p38 MAPK pathways [105]. This hypocrisy generates doubts on the definite therapeutic effect of exosomes from MSC source in various niches. In an experiment, the lack of cell response was approved in dormant-like tumor-initiating cells [116]. The differences in tumor cells to MSC secretome may relate to the divergence of factors and dynamic growth of target cells inside tumor niche [116]. In light of various genetic and proteomic reservoir, the target signaling and possible side effects of exosome treatment are required to be investigated in relation to specific distinct signaling pathway. It seems that exosome therapy is at the beginning step, and the type and source of cells have a superior role in the orientation of target cell behavior. A more deep understanding of the regulatory signaling pathways and precise inquiry in profiling of components transferred by exosomes is required to enroll and engineer the exosomes for therapeutic angiogenesis or targeted therapy

3. The application of MSCs and secretome in ischemic cardiac disease

Cardiovascular diseases remain the leading cause of mortality and morbidity in worldwide. Various investigators have continued to assess a large number of cell types injected through several routes to promote cardiac repair in patients with cardiovascular diseases in both the preclinical and clinical stages. Clinical studies have largely been focused on the administration of MSCs [117, 118]. For instance, intracoronary injection of bone marrow MSCs led to an improved function of the left ventricle in subjects with acute myocardial infarction [119]. Mechanisms of action of MSCs administrated to the injured myocardium include accelerating angiogenesis process, diminished fibrosis, and regulation of immune response [102, 120]. Both in vitro and in vivo investigations have confirmed the trans-differentiating capacity of MSCs to effective cardiomyocytes in injured cardiac tissue [50]. In addition, documents revealed that MSCs from different sources release greater amounts of angiogenic factors (HGF, VEGF, and other growth factors), cell migration chemokine (SDF-1α), immune-signaling elements (IL-6, IL-8, and

MCP-1) TGF-β, neurotrophic factors (brain-derived neurotrophic factor (BDNF)), nitric oxide (NO), and improved cardiac restoration after injury [121].

Exosomes from MSCs exposed to hypoxia and FBS-free condition enhanced neovascularization in the injured heart [92, 122–124]. In a preclinical study, intramyocardial transplantation of exosome secreted from MSCs significantly improves blood flow rate and reduced infarct zone in the rat model [125]. Approximately, the entire small and large animal model of CVD preclinical investigations along with high-quality phase 0, I, II, and III clinical trials and meta-analysis studies vigorously confirmed that MSC therapy has the effective effects in developing angiogenic networks in ischemic regions [126, 127].

Ongoing researches on preconditioning and genetic manipulations of MSCs are needed to enhance angiocrine capacity governed by growth factors, microvesicles, microRNAs, long noncoding RNAs (lncRNAs), etc. [128, 129]. Finding the route of cell delivery, the optimum dose, the excellent cell source, and transplantation time are factors that still require to be addressed so as to achieve the aim of comprehensive cardiac regeneration.

4. Angiogenesis assays

Both in vitro and in vivo angiogenesis assays are commonly used to investigate pro- and anti-angiogenic potential of stem cells and different cell types.

4.1 In vitro analyses

4.1.1 Proliferation and survival assays

Monitoring the proliferation of ECs is needed to develop microvascular units. Different survival and proliferation assays based on DNA synthesis or metabolic status are applicable. These assays could also predetermine the anti-angiogenic property of a specific compound in the context of tumor biology.

4.1.2 Migration assays

This method shows the migration in response to diverse factors, ability to digest basal membrane, and healing capacity of MSCs which is done by various assays as follows: Boyden chamber assay, Transwell® inserts, agarose assay, wound-healing assay, Teflon fence assay, phagokinetic track, etc. [130].

4.1.3 Tube formation (tubulogenesis) assay

This system is done in the 2D and 3D milieu and able to monitor alignment and juxtacrine connection of cells after plating on a specific substrate such as Matrigel, Fibrin, etc. Plated cells acquire phenotype to form capillary-like structures and lumen which are applicable to in vivo condition and evaluated in terms of tube area and number per microscopic field [130, 131].

4.1.4 Aortic ring assay

In this assay, the aorta from mouse or rats was removed and placed on collagen or fibrin matrix in serum-free condition. The angiogenic potential is determined by EC sprouting, polarized cells, and outgrowth appearance to the periphery [132].

4.2 In vivo analyses

4.2.1 Corneal angiogenesis assay

The cornea is considered as avascular tissue with unique properties for monitoring the angiogenesis and done in the model of mouse, rat, and rabbit. In the procedure of corneal angiogenesis, the candidate biomaterials and polymer with putative pro- and anti-angiogenic factors were transplanted into the stromal pouch created by surgical approach. The penetration and ingrowth of nascent vessels into the avascular area is monitored by the time [133].

4.2.2 Chicken chorioallantoic membrane angiogenesis assay

This assay is performed on embryonated eggs by using polymer pellets and silastic rings containing target molecules on the surface of the chorioallantoic membrane. After the completion of distinct time, the number and dilation of blood vessels from avian source to the implants were quantified [70].

4.2.3 Matrigel plug assay

It is a choice of in vivo angiogenesis assay following administration of gelatinous protein mixture termed Matrigel into subcutaneous space. The target molecules could be administrated with Matrigel at the site of injection and systemically to the circulation system. To precisely elucidate the formation of de novo capillaries, fluorochrome agent could be administrated into the systemic circulation [130].

5. Conclusion

It is anticipated that MSC secretome and angiocrine could be used as an off-the-shelf alternative therapy to modulate angiogenesis/vascularization in distinct tissues. Considering both pro- and anti-angiogenic capacity, a big question remains to the identification of safety and efficacy of MSC secretome under specific conditions. Based on the data from different experiments, the angiogenic paracrine potential of MSCs is currently under investigation, and results of preclinical and translational studies, if confirmatory of previous basic experiments, could lead to human medicine for angiogenic modulation of tissues. The discovery of the signaling pathways that mastermind the paracrine pro- and anti-angiogenic potential of MSCs enables us to find appropriate policies for modulating angiogenic switch on/off in in vivo condition.

Acknowledgements

All authors declared that there is no conflict of interest regarding the content of the current chapter.

Author details

Jafar Rezaie[1], Morteza Heidarzadeh[2], Mehdi Hassanpour[2,3], Hassan Amini[4], Elhameh Shokrollahi[2,3], Mahdi Ahmadi[5] and Reza Rahbarghazi[2,6]*

1 Solid Tumor Research Center, Cellular and Molecular Medicine Institute, Urmia University of Medical Sciences, Urmia, Iran

2 Stem Cell Research Center, Tabriz University of Medical Sciences, Tabriz, Iran

3 Department of Clinical Biochemistry, Tabriz University of Medical Sciences, Tabriz, Iran

4 Department of General and Vascular Surgery, Tabriz University of Medical Sciences, Tabriz, Iran

5 Department of Physiology, Faculty of Medicine, Tabriz University of Medical Sciences, Tabriz, Iran

6 Department of Applied Cell Sciences, Faculty of Advanced Medical Sciences, Tabriz University of Medical Sciences, Tabriz, Iran

*Address all correspondence to: rezarahbardvm@gmail.com; rahbarghazir@tbzmed.ac.ir

References

[1] Ucuzian AA, Gassman AA, East AT, Greisler HP. Molecular mediators of angiogenesis. Journal of Burn Care & Research. 2010;**31**(1):158-175

[2] Hou L, Kim JJ, Woo YJ, Huang NF. Stem cell-based therapies to promote angiogenesis in ischemic cardiovascular disease. American Journal of Physiology-Heart and Circulatory Physiology. 2015;**310**(4):H455-HH65

[3] Watt SM, Gullo F, van der Garde M, Markeson D, Camicia R, Khoo CP, et al. The angiogenic properties of mesenchymal stem/stromal cells and their therapeutic potential. British Medical Bulletin. 2013;**108**(1):25-53

[4] Ahmadi M, Rahbarghazi R, Aslani MR, Shahbazfar AA, Kazemi M, Keyhanmanesh R. Bone marrow mesenchymal stem cells and their conditioned media could potentially ameliorate ovalbumin-induced asthmatic changes. Biomedicine and Pharmacotherapy. 2017;**85**:28-40

[5] Keyhanmanesh R, Rahbarghazi R, Ahmadi M. Systemic transplantation of mesenchymal stem cells modulates endothelial cell adhesion molecules induced by ovalbumin in rat model of asthma. Inflammation. 2018:1-10

[6] Keyhanmanesh R, Rahbarghazi R, Aslani MR, Hassanpour M, Ahmadi M. Systemic delivery of mesenchymal stem cells condition media in repeated doses acts as magic bullets in restoring IFN-γ/IL-4 balance in asthmatic rats. Life Sciences. 2018;**212**:30-36. DOI: 10.1016/j.lfs.2018.09.049

[7] Zhao JJ, Liu JL, Liu L, Jia HY. Protection of mesenchymal stem cells on acute kidney injury. Molecular Medicine Reports. 2014;**9**(1):91-96

[8] Chen Y, Xiang LX, Shao JZ, Pan RL, Wang YX, Dong XJ, et al. Recruitment of endogenous bone marrow mesenchymal stem cells towards injured liver. Journal of Cellular and Molecular Medicine. 2010;**14**(6b):1494-1508

[9] Roubelakis MG, Tsaknakis G, Pappa KI, Anagnou NP, Watt SM. Spindle shaped human mesenchymal stem/stromal cells from amniotic fluid promote neovascularization. PLoS ONE. 2013;**8**(1):e54747-e. DOI: 10.1371/journal.pone.0054747

[10] Pankajakshan D, Agrawal DK. Mesenchymal stem cell paracrine factors in vascular repair and regeneration. Journal of Biomedical Technology and Research. 2014;**1**(1). DOI: 10.19104/jbtr.2014.107

[11] Bradshaw WJ, Rehman S, Pham TT, Thiyagarajan N, Lee RL, Subramanian V, et al. Structural insights into human angiogenin variants implicated in Parkinson's disease and amyotrophic lateral sclerosis. Scientific Reports. 2017;7:41996

[12] Souma T, Thomson BR, Heinen S, Anna Carota I, Yamaguchi S, Onay T, et al. Context-dependent functions of angiopoietin 2 are determined by the endothelial phosphatase VEPTP. Proceedings of the National Academy of Sciences. 2018;**115**(6):1298. DOI: 10.1073/pnas.1714446115

[13] Yan Z-X, Luo Y, Liu N-F. Blockade of angiopoietin-2/Tie2 signaling pathway specifically promotes inflammation-induced angiogenesis in mouse cornea. International Journal of Ophthalmology. 2017;**10**(8):1187-1194. DOI: 10.18240/ijo.2017.08.01

[14] NassiriSM,RahbarghaziR.Interactions of mesenchymal stem cells with endothelial cells. Stem Cells and Development. 2013;**23**(4):319-332

[15] Chittiboina P, Ganta V, Monceaux CP, Scott LK, Nanda A,

Alexander JS. Angiopoietins as promising biomarkers and potential therapeutic targets in brain injury. Pathophysiology. 2013;**20**(1):15-21

[16] Wang C-Q, Huang Y-W, Wang S-W, Huang Y-L, Tsai C-H, Zhao Y-M, et al. Amphiregulin enhances VEGF-A production in human chondrosarcoma cells and promotes angiogenesis by inhibiting miR-206 via FAK/c-Src/PKCδ pathway. Cancer Letters. 2017;**385**:261-270. DOI: 10.1016/j.canlet.2016.10.010

[17] Fielder GC, Yang TW-S, Razdan M, Li Y, Lu J, Perry JK, et al. The GDNF family: A role in cancer? Neoplasia (New York, NY). 2017;**20**(1):99-117. DOI: 10.1016/j.neo.2017.10.010

[18] van den Hengel LG, Versteeg HH. Tissue factor signaling: A multi-faceted function in biological processes. Frontiers in Bioscience. 2011;**3**:1500-1510

[19] Kumara HS, Pettke E, Shah A, Yan X, Cekic V, Downing MA, et al. Plasma levels of the proangiogenic protein CXCL16 remains elevated for 1 month after minimally invasive colorectal cancer resection. World Journal of Surgical Oncology. 2018;**16**(1):132

[20] Wesley UV, Hatcher JF, Ayvaci ER, Klemp A, Dempsey RJ. Regulation of dipeptidyl peptidase IV in the post-stroke rat brain and in vitro ischemia: Implications for chemokine-mediated neural progenitor cell migration and angiogenesis. Molecular Neurobiology. 2017;**54**(7):4973-4985. DOI: 10.1007/s12035-016-0039-4

[21] Kim D, Dai J, Park Y-H, Fai LY, Wang L, Pratheeshkumar P, et al. Activation of epidermal growth factor receptor/p38/hypoxia-inducible factor-1α is pivotal for angiogenesis and tumorigenesis of malignantly transformed cells induced by hexavalent chromium. Journal of Biological Chemistry. 2016;**291**(31):16271-16281. DOI: 10.1074/jbc.M116.715797

[22] Traboulsi W, Sergent F, Boufettal H, Brouillet S, Slim R, Hoffmann P, et al. Antagonism of EG-VEGF receptors as targeted therapy for choriocarcinoma progression in vitro and in vivo. Clinical Cancer Research. 2017;**23**:7130-7140. DOI: 10.1158/1078-0432.CCR-17-0811

[23] Zheng W, Aspelund A, Alitalo K. Lymphangiogenic factors, mechanisms, and applications. The Journal of Clinical Investigation. 2014;**124**(3):878-887

[24] Núñez-Gómez E, Pericacho M, Ollauri-Ibáñez C, Bernabéu C, López-Novoa JM. The role of endoglin in post-ischemic revascularization. Angiogenesis. 2017;**20**(1):1-24

[25] Xu Y, Zhu W, Wang T, Jin L, Liu T, Li X, et al. Low molecule weight fucoidan mitigates atherosclerosis in ApoE (-/-) mouse model through activating multiple signal pathway. Carbohydrate Polymers. 2019;**206**:110-120

[26] McDonnell N, Evans RM, Downes M. Methods of treating lipodystrophy using fgf-1 compounds. In: Google Patents. 2018

[27] Hosaka K, Yang Y, Nakamura M, Andersson P, Yang X, Zhang Y, et al. Dual roles of endothelial FGF-2–FGFR1–PDGF-BB and perivascular FGF-2–FGFR2–PDGFRβ signaling pathways in tumor vascular remodeling. Cell Discovery. 2018;**4**(1):3

[28] Kook S-H, Jeon Y-M, Lim S-S, Jang M-J, Cho E-S, Lee S-Y, et al. Fibroblast growth factor-4 enhances proliferation of mouse embryonic stem cells via activation of c-Jun signaling. PLoS ONE. 2013;**8**(8):e71641. DOI: 10.1371/journal.pone.0071641

[29] Chen M, Ba H, Lu C, Dai J, Sun J. Glial cell line-derived neurotrophic

factor (GDNF) promotes angiogenesis through the demethylation of the fibromodulin (FMOD) promoter in glioblastoma. Medical Science Monitor: International Medical Journal of Experimental and Clinical Research. 2018;**24**:6137

[30] Ribatti D, Tamma R. Hematopoietic growth factors and tumor angiogenesis. Cancer Letters. 2019;**440-441**:47-53. DOI: 10.1016/j.canlet.2018.10.008

[31] Dredge K, Hammond E, Handley P, Gonda T, Smith M, Vincent C, et al. PG545, a dual heparanase and angiogenesis inhibitor, induces potent anti-tumour and anti-metastatic efficacy in preclinical models. British Journal of Cancer. 2011;**104**(4):635

[32] Li Z, Yan H, Yuan J, Cao L, Lin A, Dai H, et al. Pharmacological inhibition of heparin-binding EGF-like growth factor promotes peritoneal angiogenesis in a peritoneal dialysis rat model. Clinical and Experimental Nephrology. 2018;**22**(2):257-265

[33] Van de Kamp J, Paefgen V, Wöltje M, Böbel M, Jaekel J, Rath B, et al. Mesenchymal stem cells can be recruited to wounded tissue via hepatocyte growth factor-loaded biomaterials. Journal of Tissue Engineering and Regenerative Medicine. 2017;**11**(11):2988-2998

[34] Wang C-G, Lou Y-T, Tong M-J, Zhang L-L, Zhang Z-J, Feng Y-Z, et al. Asperosaponin VI promotes angiogenesis and accelerates wound healing in rats via up-regulating HIF-1α/VEGF signaling. Acta Pharmacologica Sinica. 2018;**39**(3):393

[35] Syed SN, Jung M, Weigert A, Brüne B. S1P provokes tumor lymphangiogenesis via macrophage-derived mediators such as IL-1β or lipocalin-2. Mediators of Inflammation. 2017;**2017**

[36] Tang R, Zhang G, Chen S-Y. Smooth muscle cell proangiogenic phenotype induced by cyclopentenyl cytosine promotes endothelial cell proliferation and migration. The Journal of Biological Chemistry. 2016;**291**(52):26913-26921. DOI: 10.1074/jbc.M116.741967

[37] Shi J, Wei P-K. Interleukin-8: A potent promoter of angiogenesis in gastric cancer. Oncology Letters. 2016;**11**(2):1043-1050. DOI: 10.3892/ol.2015.4035

[38] Hubert A, Bochenek ML, Schütz E, Gogiraju R, Münzel T, Schäfer K. Selective deletion of leptin signaling in endothelial cells enhances neointima formation and phenocopies the vascular effects of diet-induced obesity in mice. Arteriosclerosis, Thrombosis, and Vascular Biology. 2017 ATVBAHA.117.309798

[39] Chen L, Liu G-Q, Wu H-Y, Jin J, Yin X, Li D, et al. Monocyte chemoattractant protein 1 and fractalkine play opposite roles in angiogenesis via recruitment of different macrophage subtypes. International Journal of Ophthalmology. 2018;**11**(2):216

[40] Fang C, Wen G, Zhang L, Lin L, Moore A, Wu S, et al. An important role of matrix metalloproteinase-8 in angiogenesis in vitro and in vivo. Cardiovascular Research. 2013;**99**(1):146-155

[41] Nielsen SH, Flynn E, Lindsey M, editors. Macrophage-derived osteopontin is fragmented by MMP-9 to hinder angiogenesis in the post-myocardial infarction left ventricle. In: ESC Congress 2017. 2017

[42] Lok J, Sardi SP, Guo S, Besancon E, Ha DM, Rosell A, et al. Neuregulin-1 signaling in brain endothelial cells. Journal of Cerebral Blood Flow and Metabolism. 2009;**29**(1):39-43. DOI: 10.1038/jcbfm.2008.94

[43] Rajkovic I, Wong R, Lemarchand E, Rivers-Auty J, Rajkovic O, Garlanda C,

et al. Pentraxin 3 promotes long-term cerebral blood flow recovery, angiogenesis, and neuronal survival after stroke. Journal of Molecular Medicine (Berlin, Germany). 2018;**96**(12):1319-1332. DOI: 10.1007/s00109-018-1698-6

[44] Liu W, Zhang J, Yao X, Jiang C, Ni P, Cheng L, et al. Bevacizumab-enhanced antitumor effect of 5-fluorouracil via upregulation of thymidine phosphorylase through vascular endothelial growth factor A/vascular endothelial growth factor receptor 2-specificity protein 1 pathway. Cancer Science. 2018;**109**(10):3294

[45] Gianni-Barrera R, Butschkau A, Uccelli A, Certelli A, Valente P, Bartolomeo M, et al. PDGF-BB regulates splitting angiogenesis in skeletal muscle by limiting VEGF-induced endothelial proliferation. Angiogenesis. 2018;**21**(4):883-900

[46] Pang V, Bates DO, Leach L. Regulation of human feto-placental endothelial barrier integrity by vascular endothelial growth factors: Competitive interplay between VEGF-A165a, VEGF-A165b, PIGF and VE-cadherin. Clinical Science. 2017. CS20171252

[47] Nakajima R, Nakamura E, Harigaya T. Vasoinhibin, an n-terminal prolactin fragment, directly inhibits cardiac angiogenesis in three-dimensional heart culture. Frontiers in Endocrinology. 2017;**8**:4

[48] Dai L, Liu Y, Xie L, Wu X, Qiu L, Di W. Sphingosine kinase 1/sphingosine-1-phosphate (S1P)/S1P receptor axis is involved in ovarian cancer angiogenesis. Oncotarget. 2017;**8**(43):74947

[49] Ma Y-L, Li W-D, Lei F-R, Qian A-M, Zhu L-W, Jiang K, et al. Metformin inhibits angiogenesis in endothelial progenitor cells through inhibiting MMP2, MMP9 and uPA expression via AMPK-mTOR-autophagy pathway. International Journal of Clinical and Experimental Medicine. 2017;**10**(1):958-964

[50] Rahbarghazi R, Nassiri SM, Ahmadi SH, Mohammadi E, Rabbani S, Araghi A, et al. Dynamic induction of pro-angiogenic milieu after transplantation of marrow-derived mesenchymal stem cells in experimental myocardial infarction. International Journal of Cardiology. 2014;**173**(3):453-466. DOI: 10.1016/j.ijcard.2014.03.008

[51] Shen C, Lie P, Miao T, Yu M, Lu Q, Feng T, et al. Conditioned medium from umbilical cord mesenchymal stem cells induces migration and angiogenesis. Molecular Medicine Reports. 2015;**12**(1):20-30

[52] Bussche L, Van de Walle GR. Peripheral blood-derived mesenchymal stromal cells promote angiogenesis via paracrine stimulation of vascular endothelial growth factor secretion in the equine model. Stem Cells Translational Medicine. 2014;**3**(12):1514-1525

[53] Kwon HM, Hur S-M, Park K-Y, Kim C-K, Kim Y-M, Kim H-S, et al. Multiple paracrine factors secreted by mesenchymal stem cells contribute to angiogenesis. Vascular Pharmacology. 2014;**63**(1):19-28

[54] Chen J-X, Zeng H, Reese J, Aschner JL, Meyrick B. Overexpression of angiopoietin-2 impairs myocardial angiogenesis and exacerbates cardiac fibrosis in the diabetic db/db mouse model. American Journal of Physiology: Heart and Circulatory Physiology. 2012;**302**(4):H1003-H1H12. DOI: 10.1152/ajpheart.00866.2011

[55] Yun J, Park S, Kim J, Park Y, Cho C. Angiopoietin 2 induces astrocyte apoptosis via αvβ5-integrin signaling in diabetic retinopathy. Cell Death & Disease. 2017;7(2):e2101

[56] Sivaraman Siveen K, Prabhu K, Krishnankutty R, Kuttikrishnan S, Tsakou M, Q Alali F, et al. Vascular endothelial growth factor (VEGF) signaling in tumour vascularization: Potential and challenges. Current Vascular Pharmacology. 2017;**15**(4):339-351

[57] Sui H, Zhao J, Zhou L, Wen H, Deng W, Li C, et al. Tanshinone IIA inhibits β-catenin/VEGF-mediated angiogenesis by targeting TGF-β1 in normoxic and HIF-1α in hypoxic microenvironments in human colorectal cancer. Cancer Letters. 2017;**403**:86-97

[58] Jian J, Pang Y, Yan HH, Min Y, Achyut BR, Hollander MC, et al. Platelet factor 4 is produced by subsets of myeloid cells in premetastatic lung and inhibits tumor metastasis. Oncotarget. 2017;**8**(17):27725

[59] Chang I-W, Liu K-W, Ragunanan M, He H-L, Shiue Y-L, Yu S-C. SERPINB5 expression: Association with CCRT response and prognostic value in rectal cancer. International Journal of Medical Sciences. 2018;**15**(4):376

[60] Meidan R, Girsh E, Mamluk R, Levy N, Farberov S. Luteolysis in ruminants: Past concepts, new insights, and persisting challenges. In: The Life Cycle of the Corpus Luteum. Springer; 2017. pp. 159-182

[61] Baxter-Holland M, Dass CR. Pigment epithelium-derived factor: A key mediator in bone homeostasis and potential for bone regenerative therapy. Journal of Pharmacy and Pharmacology. 2018;**70**(9):1127-1138

[62] Jackson HW, Defamie V, Waterhouse P, Khokha R. TIMPs: Versatile extracellular regulators in cancer. Nature Reviews Cancer. 2017;**17**(1):38

[63] Sandrim VC, Diniz S, Eleuterio NM, Gomes KB, Dusse LMS, Cavalli RC. Higher levels of circulating TIMP-4 in preeclampsia is strongly associated with clinical parameters and microRNA. Clinical and Experimental Hypertension. 2018;**40**(7):609-612

[64] Samuel SM, Ghosh S, Majeed Y, Ding H, Triggle CR. Anti-angiogenic effects of metformin in 2-deoxyglucose treated microvascular endothelial cells: Role of thrombospondin-1. The FASEB Journal. 2017;**31**(1 Supplement):824.5

[65] Nova-Lampeti E, Aguilera V, Oporto K, Guzmán P, Ormazábal V, Zúñiga F, et al. Hox genes in adult tissues and their role in endothelial cell differentiation and angiogenesis. In: Endothelial Dysfunction—Old Concepts and New Challenges. IntechOpen; 2018

[66] Nakagawa S, Okabe H, Ouchi M, Tokunaga R, Umezaki N, Higashi T, et al. Enhancer of zeste homolog 2 (EZH2) regulates tumor angiogenesis and predicts recurrence and prognosis of intrahepatic cholangiocarcinoma. HPB. 2018

[67] Liu J, Hao H, Xia L, Ti D, Huang H, Dong L, et al. Hypoxia pretreatment of bone marrow mesenchymal stem cells facilitates angiogenesis by improving the function of endothelial cells in diabetic rats with lower ischemia. PLoS ONE. 2015;**10**(5):e0126715. DOI: 10.1371/journal.pone.0126715

[68] Lobov IB, Brooks PC, Lang RA. Angiopoietin-2 displays VEGF-dependent modulation of capillary structure and endothelial cell survival in vivo. Proceedings of the National Academy of Sciences of the United States of America. 2002;**99**(17):11205-11210. DOI: 10.1073/pnas.172161899

[69] Zavala G, Prieto CP, Villanueva AA, Palma V. Sonic hedgehog (SHH) signaling improves the angiogenic potential of Wharton's jelly-derived mesenchymal stem cells (WJ-MSC). Stem Cell Research & Therapy.

2017;**8**(1):203. DOI: 10.1186/s13287-017-0653-8

[70] Matluobi D, Araghi A, Maragheh BFA, Rezabakhsh A, Soltani S, Khaksar M, et al. Carvacrol promotes angiogenic paracrine potential and endothelial differentiation of human mesenchymal stem cells at low concentrations. Microvascular Research. 2018;**115**:20-27

[71] Mandò C, Razini P, Novielli C, Anelli GM, Belicchi M, Erratico S, et al. Impaired angiogenic potential of human placental mesenchymal stromal cells in intrauterine growth restriction. Stem Cells Translational Medicine. 2016;**5**(4):451-463. DOI: 10.5966/sctm.2015-0155

[72] Zhang S, Pan C, Lv X, Wu W, Chen H, Wu W, et al. Repression of Abd-B by polycomb is critical for cell identity maintenance in adult Drosophila testis. Scientific Reports. 2017;**7**(1):5101. DOI: 10.1038/s41598-017-05359-0

[73] Vizoso FJ, Eiro N, Cid S, Schneider J, Perez-Fernandez R. Mesenchymal stem cell secretome: Toward cell-free therapeutic strategies in regenerative medicine. International Journal of Molecular Sciences. 2017;**18**(9):1852. DOI: 10.3390/ijms18091852

[74] Villar J, Carroli G, Wojdyla D, Abalos E, Giordano D, Ba'aqeel H, et al. Preeclampsia, gestational hypertension and intrauterine growth restriction, related or independent conditions? American Journal of Obstetrics and Gynecology. 2006;**194**(4):921-931

[75] Rezaie J, Mehranjani MS, Rahbarghazi R, Shariatzadeh MA. Angiogenic and restorative abilities of human mesenchymal stem cells were reduced following treatment with serum from diabetes mellitus type 2 patients. Journal of Cellular Biochemistry. 2018;**119**(1):524-535

[76] Torán JL, Aguilar S, López JA, Torroja C, Quintana JA, Santiago C, et al. CXCL6 is an important paracrine factor in the pro-angiogenic human cardiac progenitor-like cell secretome. Scientific Reports. 2017;**7**(1):12490. DOI: 10.1038/s41598-017-11976-6

[77] Nemati S, Rezabakhsh A, Khoshfetrat AB, Nourazarian A, Biray Avci Ç, Goker Bagca B, et al. Alginate-gelatin encapsulation of human endothelial cells promoted angiogenesis in in vivo and in vitro milieu. Biotechnology and Bioengineering. 2017;**114**(12):2920-2930

[78] Katare R, Riu F, Rowlinson J, Lewis A, Holden R, Meloni M, et al. Perivascular delivery of encapsulated mesenchymal stem cells improves postischemic angiogenesis via paracrine activation of VEGF-A significance. Arteriosclerosis, Thrombosis, and Vascular Biology. 2013;**33**(8):1872-1880

[79] Costa MH, McDevitt TC, Cabral JM, da Silva CL, Ferreira FC. Tridimensional configurations of human mesenchymal stem/stromal cells to enhance cell paracrine potential towards wound healing processes. Journal of Biotechnology. 2017;**262**:28-39

[80] Corrado C, Raimondo S, Chiesi A, Ciccia F, De Leo G, Alessandro R. Exosomes as intercellular signaling organelles involved in health and disease: Basic science and clinical applications. International Journal of Molecular Sciences. 2013;**14**(3):5338-5366

[81] Rezaie J, Ajezi S, Avci ÇB, Karimipour M, Geranmayeh MH, Nourazarian A, et al. Exosomes and their application in biomedical field: Difficulties and advantages. Molecular Neurobiology. 2018;**55**(4):3372-3393

[82] Hassanpour M, Cheraghi O, Brazvan B, Hiradfar A,

AghamohammadzadehN, RahbarghaziR, et al. Chronic exposure of human endothelial progenitor cells to diabetic condition abolished the regulated kinetics activity of exosomes. Iranian Journal of Pharmaceutical Research. 2018;**17**(3):1068

[83] Théry C, Zitvogel L, Amigorena S. Exosomes: Composition, biogenesis and function. Nature Reviews Immunology. 2002;**2**(8):569

[84] Rezaie J, Nejati V, Khaksar M, Oryan A, Aghamohamadzadeh N, Shariatzadeh MA, et al. Diabetic sera disrupted the normal exosome signaling pathway in human mesenchymal stem cells in vitro. Cell and Tissue Research. 2018:1-11

[85] Kowal J, Tkach M, Théry C. Biogenesis and secretion of exosomes. Current Opinion in Cell Biology. 2014;**29**:116-125

[86] Colombo M, Moita C, van Niel G, Kowal J, Vigneron J, Benaroch P, et al. Analysis of ESCRT functions in exosome biogenesis, composition and secretion highlights the heterogeneity of extracellular vesicles. Journal of Cell Science. DOI: 10.1242/jcs.128868

[87] Trajkovic K, Hsu C, Chiantia S, Rajendran L, Wenzel D, Wieland F, et al. Ceramide triggers budding of exosome vesicles into multivesicular endosomes. Science. 2008;**319**(5867):1244-1247

[88] Robbins PD, Morelli AE. Regulation of immune responses by extracellular vesicles. Nature Reviews Immunology. 2014;**14**(3):195

[89] Aoki N, Yokoyama R, Asai N, Ohki M, Ohki Y, Kusubata K, et al. Adipocyte-derived microvesicles are associated with multiple angiogenic factors and induce angiogenesis in vivo and in vitro. Endocrinology. 2010;**151**(6):2567-2576

[90] Wang L, Zhao Y, Shi S. Interplay between mesenchymal stem cells and lymphocytes: Implications for immunotherapy and tissue regeneration. Journal of Dental Research. 2012;**91**(11):1003-1010

[91] Lee J-K, Park S-R, Jung B-K, Jeon Y-K, Lee Y-S, Kim M-K, et al. Exosomes derived from mesenchymal stem cells suppress angiogenesis by down-regulating VEGF expression in breast cancer cells. PLoS ONE. 2013;**8**(12):e84256

[92] Teng X, Chen L, Chen W, Yang J, Yang Z, Shen Z. Mesenchymal stem cell-derived exosomes improve the microenvironment of infarcted myocardium contributing to angiogenesis and anti-inflammation. Cellular Physiology and Biochemistry. 2015;**37**(6):2415-2424

[93] Anderson JD, Johansson HJ, Graham CS, Vesterlund M, Pham MT, Bramlett CS, et al. Comprehensive proteomic analysis of mesenchymal stem cell exosomes reveals modulation of angiogenesis via nuclear factor-kappaB signaling. Stem Cells. 2016;**34**(3):601-613

[94] Gong M, Yu B, Wang J, Wang Y, Liu M, Paul C, et al. Mesenchymal stem cells release exosomes that transfer miRNAs to endothelial cells and promote angiogenesis. Oncotarget. 2017;**8**(28):45200

[95] Liang X, Zhang L, Wang S, Han Q, Zhao RC. Exosomes secreted by mesenchymal stem cells promote endothelial cell angiogenesis by transferring miR-125a. Journal of Cell Science. 2016;**129**(11):2182-2189

[96] Svensson KJ, Kucharzewska P, Christianson HC, Sköld S, Löfstedt T, Johansson MC, et al. Hypoxia triggers a proangiogenic pathway involving cancer cell microvesicles and PAR-2-mediated heparin-binding EGF signaling in

endothelial cells. Proceedings of the National Academy of Sciences. 2011;**108**(32):13147-13152

[97] Bian S, Zhang L, Duan L, Wang X, Min Y, Yu H. Extracellular vesicles derived from human bone marrow mesenchymal stem cells promote angiogenesis in a rat myocardial infarction model. Journal of Molecular Medicine. 2014;**92**(4):387-397

[98] Chen Z, Wang H, Xia Y, Yan F, Lu Y. Therapeutic potential of mesenchymal cell-derived miRNA-150-5p-expressing exosomes in rheumatoid arthritis mediated by the modulation of MMP14 and VEGF. The Journal of Immunology. 2018;**201**(8):2472-2482

[99] Shabbir A, Cox A, Rodriguez-Menocal L, Salgado M, Badiavas EV. Mesenchymal stem cell exosomes induce proliferation and migration of normal and chronic wound fibroblasts, and enhance angiogenesis in vitro. Stem Cells and Development. 2015;**24**(14):1635-1647

[100] Zou X, Gu D, Xing X, Cheng Z, Gong D, Zhang G, et al. Human mesenchymal stromal cell-derived extracellular vesicles alleviate renal ischemic reperfusion injury and enhance angiogenesis in rats. American Journal of Translational Research. 2016;**8**(10):4289

[101] Zhang J, Guan J, Niu X, Hu G, Guo S, Li Q, et al. Exosomes released from human induced pluripotent stem cells-derived MSCs facilitate cutaneous wound healing by promoting collagen synthesis and angiogenesis. Journal of Translational Medicine. 2015;**13**(1):49

[102] Lai RC, Arslan F, Lee MM, Sze NSK, Choo A, Chen TS, et al. Exosome secreted by MSC reduces myocardial ischemia/reperfusion injury. Stem Cell Research. 2010;**4**(3):214-222

[103] Camussi G, Deregibus MC, Tetta C. Paracrine/endocrine mechanism of stem cells on kidney repair: Role of microvesicle-mediated transfer of genetic information. Current Opinion in Nephrology and Hypertension. 2010;**19**(1):7-12

[104] Quesenberry PJ, Dooner MS, Aliotta JM. Stem cell plasticity revisited: The continuum marrow model and phenotypic changes mediated by microvesicles. Experimental Hematology. 2010;**38**(7):581-592

[105] Zhu W, Huang L, Li Y, Zhang X, Gu J, Yan Y, et al. Exosomes derived from human bone marrow mesenchymal stem cells promote tumor growth in vivo. Cancer Letters. 2012;**315**(1):28-37

[106] López-Cobo S, Campos-Silva C, Moyano A, Oliveira-Rodríguez M, Paschen A, Yáñez-Mó M, et al. Immunoassays for scarce tumour-antigens in exosomes: Detection of the human NKG2D-ligand, MICA, in tetraspanin-containing nanovesicles from melanoma. Journal of Nanobiotechnology. 2018;**16**(1):47

[107] Xue C, Shen Y, Li X, Li B, Zhao S, Gu J, et al. Exosomes derived from hypoxia-treated human adipose mesenchymal stem cells enhance angiogenesis through the PKA signaling pathway. Stem Cells and Development. 2018;**27**(7):456-465

[108] Yu B, Gong M, Wang Y, Millard RW, Pasha Z, Yang Y, et al. Cardiomyocyte protection by GATA-4 gene engineered mesenchymal stem cells is partially mediated by translocation of miR-221 in microvesicles. PLoS ONE. 2013;**8**(8):e73304

[109] Keshtkar S, Azarpira N, Ghahremani MH. Mesenchymal stem cell-derived extracellular vesicles: Novel frontiers in regenerative medicine. Stem Cell Research & Therapy. 2018;**9**(1):63

[110] Zhou J, Tan X, Tan Y, Li Q, Ma J, Wang G. Mesenchymal stem cell derived exosomes in cancer progression, metastasis and drug delivery: A comprehensive review. Journal of Cancer. 2018;**9**(17):3129

[111] Goodarzi P, Larijani B, Alavi-Moghadam S, Tayanloo-Beik A, Mohamadi-Jahani F, Ranjbaran N et al. Mesenchymal Stem Cells-Derived Exosomes for Wound Regeneration. 2018

[112] Rani S, Ritter T. The Exosome—A Naturally Secreted Nanoparticle and its Application to Wound Healing. Advanced Materials. 2016;**28**(27):5542-5552

[113] Zhang B, Wang M, Gong A, Zhang X, Wu X, Zhu Y, et al. HucMSC-exosome mediated-Wnt4 signaling is required for cutaneous wound healing. Stem Cells. 2015;**33**(7):2158-2168

[114] Yuan Y, Du W, Liu J, Ma W, Zhang L, Du Z, et al. Stem cell-derived exosome in cardiovascular diseases: Macro roles of micro particles. Frontiers in Pharmacology. 2018;**9**

[115] Sharma A. Role of stem cell derived exosomes in tumor biology. International Journal of Cancer. 2018;**142**(6):1086-1092

[116] Zimmerlin L, Park TS, Zambidis ET, Donnenberg VS, Donnenberg AD. Mesenchymal stem cell secretome and regenerative therapy after cancer. Biochimie. 2013;**95**(12):2235-2245. DOI: 10.1016/j.biochi.2013.05.010

[117] Jeevanantham V, Butler M, Saad A, Abdel-Latif A, Zuba-Surma EK, Dawn B. Adult bone marrow cell therapy improves survival and induces long-term improvement in cardiac parameters: A systematic review and meta-analysis. Circulation. 2012 CIRCULATIONAHA.111.086074

[118] Jeevanantham V, Afzal MR, Zuba-Surma EK, Dawn B. Clinical trials of cardiac repair with adult bone marrow-derived cells. Cellular Cardiomyoplasty. 2013:179-205

[119] Schächinger V, Erbs S, Elsässer A, Haberbosch W, Hambrecht R, Hölschermann H, et al. Intracoronary bone marrow–derived progenitor cells in acute myocardial infarction. New England Journal of Medicine. 2006;**355**(12):1210-1221

[120] Arslan F, Lai RC, Smeets MB, Akeroyd L, Choo A, Aguor EN, et al. Mesenchymal stem cell-derived exosomes increase ATP levels, decrease oxidative stress and activate PI3K/Akt pathway to enhance myocardial viability and prevent adverse remodeling after myocardial ischemia/reperfusion injury. Stem Cell Research. 2013;**10**(3):301-312

[121] Singh A, Singh A, Sen D. Mesenchymal stem cells in cardiac regeneration: A detailed progress report of the last 6 years (2010 2015). Stem Cell Research & Therapy. 2016;7(1):82. DOI: 10.1186/s13287-016-0341-0

[122] Hodgkinson CP, Bareja A, Gomez JA, Dzau VJ. Emerging concepts in paracrine mechanisms in regenerative cardiovascular medicine and biology. Circulation Research. 2016;**118**(1):95-107

[123] Huang L, Ma W, Ma Y, Feng D, Chen H, Cai B. Exosomes in mesenchymal stem cells, a new therapeutic strategy for cardiovascular diseases? International Journal of Biological Sciences. 2015;**11**(2):238

[124] Caporali A, Bäck M, Daemen MJ, Hoefer IE, Jones EA, Lutgens E, et al. Future directions for therapeutic strategies in post-ischaemic vascularization: A position paper from European Society of Cardiology Working Group on Atherosclerosis and Vascular Biology. Cardiovascular Research. 2018;**114**(11):1411-1421

[125] Lou G, Chen Z, Zheng M, Liu Y. Mesenchymal stem cell-derived exosomes as a new therapeutic strategy for liver diseases. Experimental & Molecular Medicine. 2017;**49**(6): e346-e34e. DOI: 10.1038/emm.2017.63

[126] Yong KW, Choi JR, Mohammadi M, Mitha AP, Sanati-Nezhad A, Sen A. Mesenchymal stem cell therapy for ischemic tissues. Stem Cells International. 2018;**2018**:8179075. DOI: 10.1155/2018/8179075

[127] Tao H, Han Z, Han ZC, Li Z. Proangiogenic features of mesenchymal stem cells and their therapeutic applications. Stem Cells International. 2016;**2016**

[128] Kir D, Schnettler E, Modi S, Ramakrishnan S. Regulation of angiogenesis by microRNAs in cardiovascular diseases. Angiogenesis. 2018:1-12

[129] Lemcke H, Voronina N, Steinhoff G, David R. Recent progress in stem cell modification for cardiac regeneration. Stem Cells International. 2018;**2018**

[130] Rahbarghazi R, Nassiri SM, Khazraiinia P, Kajbafzadeh A-M, Ahmadi SH, Mohammadi E, et al. Juxtacrine and paracrine interactions of rat marrow-derived mesenchymal stem cells, muscle-derived satellite cells, and neonatal cardiomyocytes with endothelial cells in angiogenesis dynamics. Stem Cells and Development. 2012;**22**(6):855-865

[131] Rezabakhsh A, Nabat E, Yousefi M, Montazersaheb S, Cheraghi O, Mehdizadeh A, et al. Endothelial cells' biophysical, biochemical, and chromosomal aberrancies in high-glucose condition within the diabetic range. Cell Biochemistry and Function. 2017;**35**(2):83-97

[132] Kruger EA, Duray PH, Tsokos MG, Venzon DJ, Libutti SK, Dixon SC, et al. Endostatin inhibits microvessel formation in the ex vivo rat aortic ring angiogenesis assay. Biochemical and Biophysical Research Communications. 2000;**268**(1):183-191

[133] Grogan JA, Connor AJ, Pitt-Francis JM, Maini PK, Byrne HM. The importance of geometry in the corneal micropocket angiogenesis assay. PLoS Computational Biology. 2018;**14**(3):e1006049

Chapter 7

The Dual Nature of Mesenchymal Stem Cells (MSCs): Yin and Yang of the Inflammatory Process

Carmen Ciavarella and Gianandrea Pasquinelli

Abstract

The well-known reparative properties of mesenchymal stem cells (MSCs) make them an attractive source for cell-based therapy. In vitro and in vivo studies support an anti-inflammatory role of MSCs by directly targeting immune cells or via the secretion of immunomodulatory factors. MSCs have been isolated from several human normal tissues, even from pathological biopsies and blood samples; in these cases, MSCs displayed peculiar characteristics, suggesting a phenotype transition into a pathological state. Indeed, MSCs derived from inflamed tissues acquired a pro-inflammatory behaviour. In this view, MSCs may be crucial players of many pathways involved in human diseases, especially during the inflammatory cascade. The present chapter will minutely describe the basic biology of human MSCs derived from normal and pathological arteries, focusing on their dual nature as cellular switchers of the inflammatory setting. We will also discuss the emerging role of miRNAs in regulating MSC functions and their potential use as alternative strategies to manipulate MSC efficacy.

Keywords: mesenchymal stem cells (MSCs), immunomodulation, inflammatory process, phenotype switching, vascular wall

1. Introduction

Mesenchymal stem cells (MSCs) are adult, multipotent stem cells endowed of self-renewal, a process of continuous divisions essential to maintain the stem cell pool. Meanwhile, MSCs can be activated under the action of growth factors, chemokines and cytokines which are normally released during the physiological tissue renewal or in pathological conditions in the presence of tissue damage. Specific signals stimulate the MSC migration at the damaged site and their differentiation into specialized cell types belonging to the mesodermal lineage. The homing and the differentiation potential allow MSCs to be actively involved in the tissue homeostasis as well as in the repair process. MSCs have been firstly identified as a non-hematopoietic, adherent and spindle-shaped cell subset of the human bone marrow stroma [1]. From their first isolation in 1970, MSCs have been extensively characterized, gaining the increasing interest of the scientific community, and lots of studies issued the biology and the inner properties of these promising cells. One of the most intriguing and studied functions of MSCs is the immunomodulation, that is, the ability to repress inflammation; however, little is known about the reversal of this property that has been observed in some disease models. The present chapter will review the differentiation and immunomodulatory capabilities of MSCs, will discuss the contradictory face of MSCs and will focus on the vascular wall setting.

2. Mesenchymal stem cells (MSCs): source, phenotype and properties

In vivo and in vitro data have demonstrated the unique reparative potential of MSCs, which are now considered as the most attractive and functional source for cell-based therapies in the field of regenerative medicine. First of all, the ease of MSC isolation and in vitro propagation excited researchers, who addressed their efforts to find novel sources for MSCs and to characterize them. MSCs have been efficiently isolated and characterized from different human tissues, other than their native site, i.e. peripheral blood [2], umbilical cord (UC) blood [3], fat [4], Wharton's jelly (WJ) [5], synovial membrane [6] and vascular wall [7, 8]. The lack of standard markers, differences in laboratory procedures, type and age of the source tissue may affect the purity of MSC pool and impair their effectiveness for clinical applications. In order to fill this gap and provide a consensus statement for MSC definition, the International Society for Cellular Therapy (ISCT) postulated the following minimal criteria [9]:

- Adherence to plastic substrate in vitro
- Expression of surface markers CD90, CD105 and CD73 (mesenchymal linage) and lack of CD34, CD19, CD45 and CD11a (hematopoietic lineage), CD31 (endothelial lineage) and HLA-DR (human leukocyte antigen)
- Multilineage differentiation ability into the mesodermal lineage (chondrogenic, adipogenic, osteogenic commitments)

Even if researchers concur with the consideration of CD34 as distinctive marker of endothelial and hematopoietic cells, it has been detected also in MSCs correlating with advanced progenitor properties [10, 11]. Moreover, adipose tissue-derived MSCs positive to CD34 lost its expression after in vitro propagation suggesting that the absence of CD34 may be a result of in vitro culture [12, 13]. Thus, the immunophenotype is not per se sufficient to identify MSCs, but functional assays aimed at testing the stemness properties, like the ability to form colonies, and the differentiation potential are necessary. At this regard, MSCs also manifest the capacity to differentiate into ectoderm- and endoderm-derived cell types, i.e. endothelial cells, neurons and hepatocytes.

The presence of MSCs with self-renewal and multilineage differentiation properties within adult tissues suggests their intrinsic participation to the regular tissue homeostasis and cell turnover. Cytokines and chemokines that are released from the injured tissue act as recruiting factors of MSCs from their niche, allowing their mobilization and trafficking. The ability of differentiate into tissue-specific cell types is the major mechanism through which MSCs replace dead cells; in addition, MSCs secrete soluble factors that include hepatocyte growth factor (HGF), transforming growth factor (TGF) β1 and vascular endothelial growth factor (VEGF) and contribute to cutaneous wound healing [14].

The migration and the differentiation abilities support the clinical use of MSCs for the cure of degenerative diseases but, if uncontrolled or impaired, could become prerequisites to the occurrence of pathological conditions. Further, the tissue source constitutes a discriminating factor among MSCs in terms of differentiation potency, migration and effectiveness in tissue repair. The differentiation potency can also be regulated on epigenetic basis; at this regard, the methylation status of the main regulators of MSC fate crucially drives the differentiation program. This condition has been demonstrated in a study by Xu et al., where BM-MSCs resulted more effective in osteogenic differentiation than adipose tissue (AT)-MSCs, which displayed the opposite trend and were mainly addressed towards the adipogenic commitment [15]. Similarly, perivascular MSCs isolated from the UC blood exhibited higher angiogenic potency than umbilical artery and WJ- MSCs, with implications for the cure of ischemic injury [16]. Therefore, the clinical use of MSCs is a promising and undeniable chance for regenerative medicine, but it needs to be optimized because of the MSC multifaceted nature

2.1 MSC immunomodulation

A large body of experimental and clinical studies showed that MSCs modulate the immune response, both innate and adaptive with possible implications in the management of transplantation, autoimmune and inflammatory disorders [17]. MSCs have been historically considered as immune-privileged cells, because of their poor immunogenicity. Indeed, the low levels of human leukocyte antigen (HLA) histocompatibility complex-I and the lack the complex HLA-II allow MSCs to elicit the immune recognition. Additionally, MSCs do not express the Fas ligand and the co-stimulatory molecules CD40, CD80 (B7-1) and CD86 (B7-2), which are necessary for effector T-cell activation [18, 19]. On the other hand, it has been widely demonstrated that MSCs affect the immune system, both through cell–cell interactions and by the paracrine secretion of anti-inflammatory factors.

2.2 MSC-immune cell interactions and in vitro immunomodulatory assays

The majority of studies use a mixed lymphocyte reaction (MLR) assay and explore the immunosuppressive effects of MSCs on allogeneic T cells in the MLR reaction. Peripheral blood mononuclear cells (PBMCs) are obtained by density gradient separation and cultured on a feeder layer of irradiated MSCs generally for 3 or 5 days. PBMCs are stimulated by the addition of mitogens to the culture media like phytohaemagglutinin (PHA), which markedly induces CD8 T-cell proliferation [20]. The coculture protocol can be executed in a direct manner for testing effects mediated by physical cell-cell interactions; alternatively, it is possible to evaluate the influence of MSC paracrine secretion in a separate coculture system by the use of a Transwell insert of 0.4 μm that inhibits PBMC migration and maintains the two cell compartments separated. At the end of the experimental coculture, PBMCs are collected and analyzed in terms of proliferation and activation. Cell cycle analysis and specific proliferation test, like the incorporation rate of bromodeoxyuridine (BrdU) into DNA, are executed for T-cell survival together with the analysis of T-cell subpopulation percentage. On the other hand, MSCs are characterized for immunomodulatory markers, such as HLA-G, interleukin (IL)-10 and HGF, through a wide range of techniques including flow cytometry, immunofluorescence and Western blot. A pictorial description of the basic immunomodulatory assay can be observed in **Figure 1a**. MSCs inhibit naïve $CD4^+$ T helper cell proliferation by inducing the cell cycle arrest at G0 and by hindering the T helper cell differentiation into T_H1 and T_H17 subsets. In addition to direct cell-cell interactions, the MSC immunomodulation is exerted through the secretion of soluble anti-inflammatory mediators, such as nitric oxide (NO), TGF-β1, interleukin (IL)-10, indoleamine 2,3-dioxygenase (IDO), HGF and prostaglandin E2 (PGE2). HLA-G is another crucial component involved into the MSC immunosuppressive system, allowing the induction of regulatory T cells (Treg) and inhibiting natural killer (NK) cell cytotoxicity and

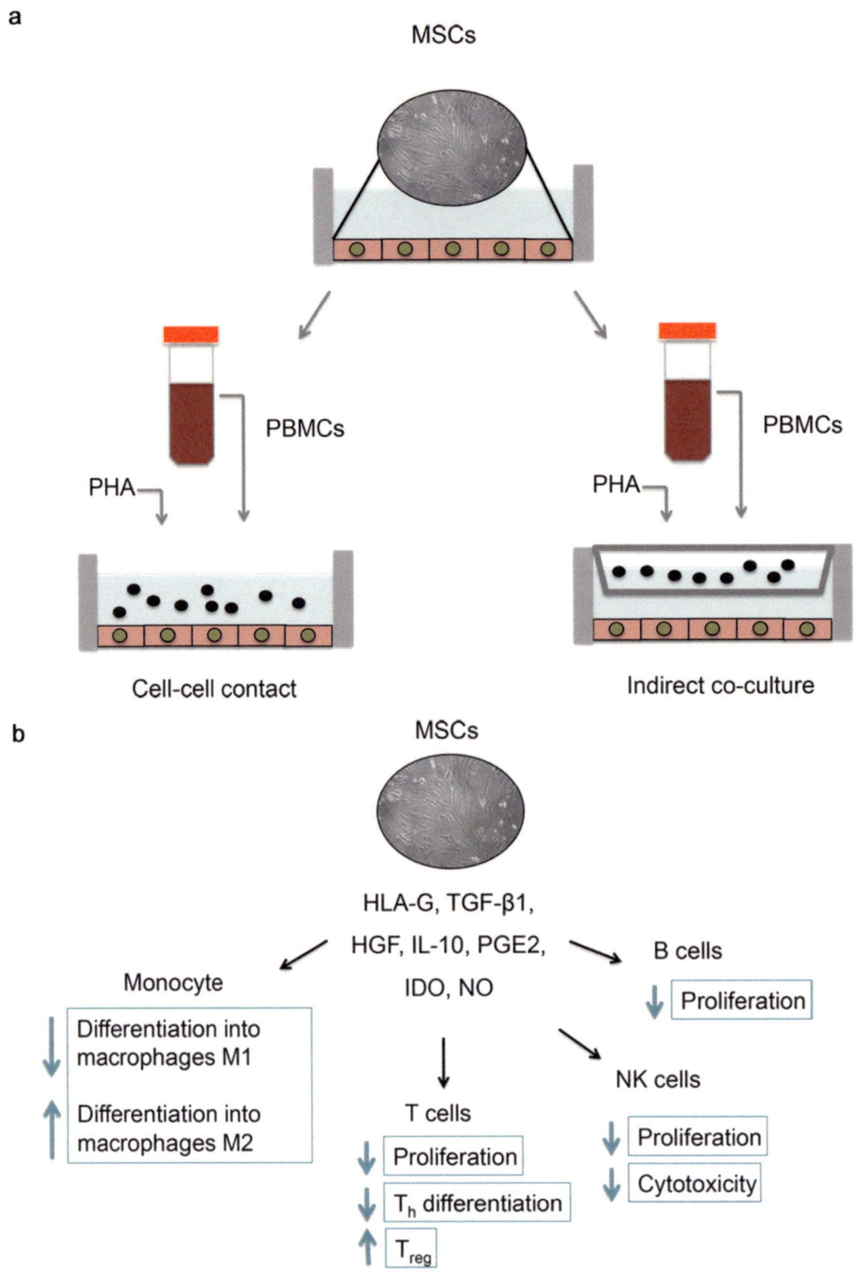

Figure 1.
MSC immunomodulation: (a) schematic description of the immunomodulatory procedure for testing the MSC effects on PBMCs; and (b) overview of the main mechanisms MSC-dependent on immune cells. Abbreviations: PBMC, peripheral blood mononuclear cells; PHA, phytohaemagglutinin; HLA-G, human leukocyte antigen; HGF, hepatocyte growth factor; TGF-β1, transforming growth factor; IL-10, interleukin; NO, nitric oxide; IDO, indoleamine 2,3-dioxygenase; PGE2, prostaglandin E2.

dendritic cell (DC) maturation [21, 22]. MSCs further exert their immunosuppressive effects on B cells, blocking their proliferation. The scheme in **Figure 1b** summarizes the main mechanisms relative to the MSC-driven immunomodulation.

The successful application of MSCs has been reported in several clinical studies; among these, MSCs resulted effective in graft-versus-host disease (GvHD) patients even in those resistant to steroid treatment. Moreover, beneficial effects were observed in patients affected by systemic lupus erythematous (SLE) and Crohn's disease [23].

However, the effectiveness and safety of MSC application are still under debate, and some data suggest that the immunomodulatory function is tightly regulated. Moreover, many studies have elucidated differences in terms of immunomodulatory potency among MSCs isolated from distinct sources. At this regard, a study by Mattar et al. compared the immunosuppressive properties of BM-MSCs, AT-MSCs, CB-MSCs and WJ-MSCs on PHA-activated T cells, showing that MSCs from tissues alternative to bone marrow were effective in inhibiting T-cell proliferation [24]. Similarly, Li et al. demonstrated that WJ-MSCs exhibited the highest immunosuppressive force and lowest levels of immunogenic factors than AT-, BM- and PL-MSCs, thus resulting as the most promising for potential therapeutic application [25, 26]. Further comparative analysis aimed at deepen functional characteristics of MSCs from multiple sources is necessary to improve their translation into the clinic.

2.3 MSCs like macrophages: switch from pro-inflammatory to anti-inflammatory phenotype

Several studies have addressed the intrinsic mechanisms associated with MSC immunomodulation, and increasingly evidences have demonstrated the plastic nature of this intriguing property. The effectiveness of MSC immunomodulation can depend on the external microenvironment; different studies support the hypothesis that MSCs can both reduce and strengthen the inflammatory process but is the inflammatory context itself at determining the immunosuppressive function of MSCs. Thus, MSCs need to be 'licensed' for their activation and regulate the immune response [23]. At this regard, MSCs can switch from a pro-inflammatory to an anti-inflammatory phenotype, characterized by a different soluble factor and cytokine panel. This paradigm mimics the macrophage polarization model. Macrophages are immune cells that derive from the differentiation of their precursor cells, monocytes, and represent key players of the immune response. Monocytes, once arrived at the damaged site, can differentiate into M1- or M2-type macrophage, depending on the microenvironment stimuli. The M1 macrophage releases a series of inflammatory cytokines, contributing to the local inflammation; conversely, the M2 macrophage mainly secretes IL-10 and TGF-β1 that are anti-inflammatory. The local microenvironment critically triggers MSC polarization, and the Toll-like receptors (TLRs) are of crucial importance in this regulatory mechanism. TLRs represent a conserved family of pattern recognition receptors (PRRs) able to detect a wide spectrum of pathogen-associated patterns (PAMPs) and activate the immune cells. The expression of TLRs in MSCs changes according to the tissue source but also depends on the microenvironment stimuli. According to Bunnel et al., the activation of TLR4 triggers MSCs skew into a pro-inflammatory phenotype, releasing factors like IL-6 and IL-8 that contribute to tissue injury [27]. Conversely, the stimulation of TLR3 shifts MSCs to the anti-inflammatory phenotype accompanied by the secretion of IL-4, IDO and PGE-2 [27, 28]. Other studies support the contribution of nitric oxide (NO) to the anti-inflammatory activity of MSCs; indeed, the inhibition or the genetic ablation of NO synthase (iNOS) stimulates T-cell proliferation after being cultured with MSCs [29]. Thus, the poor activation of MSC anti-inflammatory profile leads to the worsening of the inflammatory process and of the tissue damage. Defects of the immunomodulatory functions have been observed in some diseases, like a study on a mouse model of collagen-induced arthritis (CIA) demonstrated. MSCs exposed to an inflammatory environment, like in the CIA model, displayed impaired immunomodulation and, after TNF-α addition, lost the ability to regulate T-cell proliferation [30]. Also, MSCs isolated from the bone marrow of multiple myeloma patients were ineffective in vitro, reflecting an aberrant T-cell function in vivo [31]. The therapeutic application of MSCs for treating inflammatory conditions is really promising; nevertheless, the effectiveness and safety of their administration in vivo are still controversial, representing a challenging issue. The diversity of MSC biology and mechanism of action implicate knowledge gaps that need to be filled for a clinical application of MSCs on large scale [32, 33]. For this reason, novel studies should address all the technical concerns related to the use of MSCs and extend their investigation field to MSCs from human disease models.

3. Tissue specificity of MSCs: normal versus diseased arteries

As described in the second paragraph, MSCs are markedly represented in several adult human tissues other than bone marrow and fat; the rationale to this wide distribution comes from the existence of the vasculogenic zone, which is a stem cell niche within the vascular wall.

In this field, our research team has reached a 10-year experience in isolating MSCs from a broad range of vascular segments of small and large size, including healthy and diseased arteries. As regards the technical concerns, the isolation protocol consists of the enzymatic digestion of both fresh and cryopreserved arteries collected at the Cardiovascular Tissue and Cord Blood Bank (S. Orsola-Malpighi Hospital, Bologna, Emilia Romagna-Italy) and belonged to different vascular districts (carotid bifurcation, aortic arch, abdominal/thoracic aorta, femoral artery). In spite of the extreme temperature conservation in liquid nitrogen for about 10 years, these vascular tissues were vital and resulted as an unusual source of clonogenic and highly proliferative MSCs endowed of chondrogenic, adipogenic, osteogenic and smooth muscle differentiation capability [34]. Interestingly, these human vascular wall-MSCs (hVW-MSCs) possess angiogenic potential, as demonstrated by the capacity of forming a capillary-like network when seeded onto a semi-solid matrix (Matrigel) and by the expression of endothelial markers after VEGF stimulation. The capacity of VW-MSCs to differentiate into endothelial-like cells befits with their location within the arterial wall, suggesting their function as cell reservoir during the normal tissue renewal. The angiogenic potential and the migratory capacity of VW-MSCs were explored as crucial aspects for the healing of vascular ulcers and resulted boosted under the effects of recombinant HGF [35]. Another intriguing finding supporting the therapeutic force of VW-MSCs regarded the immunomodulatory capability. The vessel wall is prone to inflammatory infiltration following the endothelial dysfunction; thus, it is reasonable to hypothesize that vascular resident progenitors are able to contain this phenomenon. hVW-MSCs exerted a significant suppressive effect on PHA-PBMCs proliferation, partially mediated by the expression of HLA-G. These data support the existence of MSC cells within the vascular wall and their participation in the normal tissue homeostasis as well as in the arterial repair. During the early phase of the atherogenic process, monocytes cross the endothelial barrier and differentiate into macrophages within the media where the release of matrix metalloproteinases (MMPs), cytokines and chemokines recruits other monocytes and lymphocytes. This vascular inflammation is the licencing factor that activates vascular MSCs into the immunosuppressive phenotype; any intimate disturbance to this mechanism may skew MSCs into the opposite trend, failing in vascular healing.

3.1 Functional deregulation of MSCs within the injured artery

As reported in some disease models, like osteo-chondrogenic disorders or hematological affections, MSCs can undergo a deregulation of their reparative properties that can be enhanced or defected with undeniable pathological implications. Our team investigated the contradictory face of MSCs, extending the analysis to the field of cardiovascular diseases like abdominal aortic aneurysm (AAA), atherosclerotic carotid plaque and arteriovenous fistula. AAA is a chronic dilatation of the aortic wall, due to the exaggerated degradation of extracellular matrix proteins by MMPs resulting in the loss of elastin and collagen; in addition, the inflammatory infiltrate occurs and contributes to the degeneration of the media tunica, together with further release of MMPs and the destabilization of the wall. Published data have shown for the first time that, in spite of the complete disorganization of the tissue, MSCs isolated from the AAA wall (named AAA-MSCs) were comparable to the healthy counterpart in morphology, growth rate and immunophenotype [36]. Even more, these pathological MSCs displayed altered functions in vitro; consistent with the chronic inflammatory setting of the AAA disease, AAA-MSCs displayed low HLA-G expression and resulted ineffective in modulating the PBMC proliferation. This low efficacy may reflect the switch of vascular MSCs into the pro-inflammatory phenotype and the worsening of the aortic wall conditions. Angiogenesis represents another distinctive hallmark of AAA pathology, significantly affecting the wall stability. AAA-MSCs were able to differentiate into endothelial-like cells, as demonstrated by the formation of a vascular network onto Matrigel and by the positivity to CD31 marker. Nevertheless, we observed a reduced expression of CD146, a pericyte marker, suggesting the instability and immaturity of the AAA-MSC-derived neo-vessels [37]. A representative image of AAA-MSC functional characteristics is reported in **Figure 2**.

Other few works have issued the pathological role of MSCs residing within the vasculogenic zone. At this regard, the first hypothesis on MSC contribution to vascular diseases was proposed in 2012 by Tang et al., who showed that rodent MSCs undergo increased proliferation and migration upon vascular injury [38]. In 2016, a study of Kramann et al. performed on ApoE−/− mice with chronic kidney disease demonstrated the involvement of a population of MSC-like cells to the onset of athero- and arteriosclerosis as well as to the differentiation into osteoblast-like cells [39]. An aberrant differentiation program of MSCs can be crucial in triggering the complications of the atherosclerotic plaque, like ectopic bone formation,

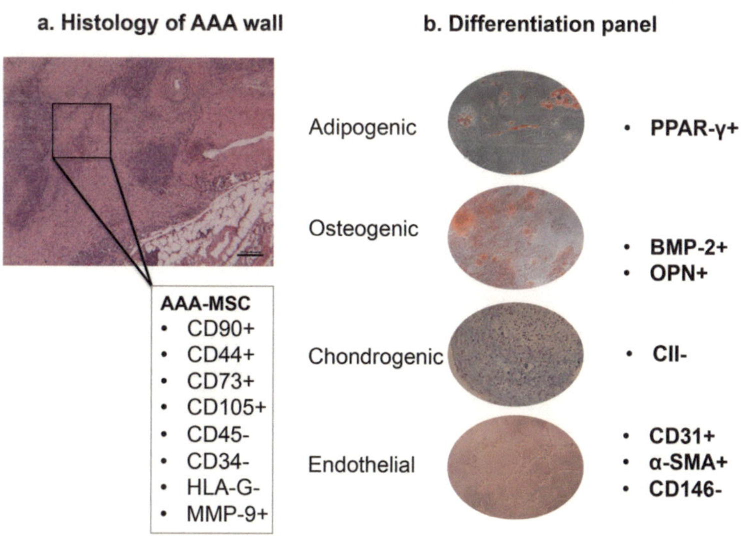

Figure 2.
Immunophenotype and differentiation assays of AAA-MSCs: (a) AAA-MSCs positive to CD90, CD44, CD73, CD105 and MMP-9 and negative to CD45 and CD34 were isolated from the aorta affected by aneurysm, inflammation and atherosclerotic plaque. (b) AAA-MSCs were characterized for adipogenic, osteogenic, chondrogenic and endothelial differentiation abilities. Abbreviations: HLA, human leukocyte antigen; MMP, matrix metalloproteinase; PPAR, peroxisome proliferator-activated receptor; BMP, bone morphogenetic protein; OPN, osteopontin; CII, collagen type 2; SMA, smooth muscle actin.

Properties	**Source**	
	Healthy arteries	Pathological aorta
Morphology	Fibroblast-like	Fibroblast-like
Immunophenotype	CD90+, CD105+, CD73+, CD44+, CD34-	CD90+, CD105+, CD73+, CD44+, CD34-
Multilineage differentiation		
Adipogenic	++	++
Osteogenic	+	+++
Chondrogenic	++	-
Angiogenic potency	+	++
	CD31+, α-SMA+, CD146+	CD31+, α-SMA+, CD146-
Immunomodulation		
T cell proliferation	++	-
HLA-G expression	+	-

Table 1.
Phenotypic and functional characteristics of healthy versus pathological vascular MSCs.

and represents the early stage during calcification process [40]. AAA-MSCs also exhibited a marked osteogenic ability, correlating with the vascular calcium levels as measured by angio-CT in the enrolled patients [37]. Thus, it can be postulated that MSCs are key players during the renewal as well as the pathological conditions affecting the vascular wall. The MSC behaviour can be seen as fine balance between two opposite forces, which is strongly influenced by the external microenvironment and the interaction with the neighboring cells. At this regard, the immune cells and the cytokines released during inflammation are key factors in exacerbating the osteogenic differentiation of healthy VW-MSCs [37].

The most remarkable characteristics of AAA-MSCs and their comparison to the hVW-MSCs are summarized in **Table 1**.

4. miRNA regulation of MSC immunomodulatory capacity

Micro-RNAs (miRNAs) constitute a class of single-stranded non-coding RNAs of approximately 18–22 nucleotides that function as endogenous regulators of gene expression through the degradation of the target mRNA or the inhibition of the transcription process. Over the last decade, the growing interest for miRNA applications elucidated their involvement into many biological mechanisms, like cell growth and proliferation. Based on the observation that miRNAs are differentially regulated in the presence of pathological conditions like cancer or immune diseases, many researchers have proposed their use as diagnostic markers or therapeutic targets. Several studies revealed the miRNA involvement into the hematopoietic stem cell (HSC) system, driving aspects like cell survival, self-renewal and differentiation. Moreover, these 'immuno-miRs' orchestrate crucial steps of both innate and adaptive immune cell development and function [41]. miR-21, miR-146a and miR-155 are included in this category and are induced upon T-cell receptor (TCR) activation through the NF-kB cascade. miRNAs have also been shown to regulate the stem cell behaviour, self-renewal and differentiation; therefore, investigating immuno-miRs in human MSCs could be suggestive of their reparative properties like differentiation and immunomodulatory potency. Some of the described immuno-miRs target the TLR pathway in MSCs or immune cells suppressing or enhancing TLR activation by targeting adapter molecules, cytokines and transcription factors. At this regard, a work by Matysiak et al. demonstrated the upregulation of several miRNAs in differentiated BMSCs that had lost immunomodulation, including miR-146a, together with a low expression of PGE2 [42]. The role of miR-155 in interfering with MSC immunomodulation has been also reported; indeed, miR-155 decreases the iNOS production in cytokine-activated MSCs, partially targeting TGF-beta-activated kinase 2 (TAB2), an adapter protein involved in TLR pathways [43]. As recently demonstrated, many of the therapeutic effects of MSCs are mediated by the extracellular vesicles (EVs), which are membrane-bound vesicles that serve as vehicle of mRNAs and proteins. Moreover, EVs are enriched of miRNAs that are released in the circulatory system. The EV-released miRNAs, such as Let-7b, miR-1180 and miR-183, induce macrophage polarization into M2 phenotype and mitigate inflammation by reducing TLR4 [44]. The mechanisms regulating the miRNA-MSCs interplay are complex and require further investigations; the use of miRNA-enriched EVs derived from human MSCs could be the promising therapeutic cell-free alternative for the cure of GvHD and inflammatory diseases.

5. Conclusions

The regenerative and reparative properties of MSCs are certainly undisguised even though many efforts are necessary to ensure their use for clinical therapy. The contradictory inflammatory activity of MSCs is a result of their plastic nature and represents a critical issue that needs to be addressed. Tissue-resident MSCs can represent the optimal target of stem cell-reprogramming therapies aimed at restoring their native reparative properties. The emerging role of miRNAs in regulating the MSC functions is promising and requires further investigations for miRNA manipulation in order to address MSC towards a more efficient and safe reparative activity.

Author details

Carmen Ciavarella* and Gianandrea Pasquinelli
Clinical Pathology, Department of Experimental, Diagnostic and Specialty Medicine, (DIMES)-University of Bologna, S. Orsola-Malpighi University Hospital of Bologna, Bologna, Italy

*Address all correspondence to: carmen.ciavarella2@unibo.it

References

[1] Friedenstein AJ, Chailakhjan RK, Lalykina KS. The development of fibroblast colonies in monolayer cultures of guinea-pig bone marrow and spleen cells. Cell and Tissue Kinetics. 1970;**3**(4):393-403

[2] Zvaifler NJ, Marinova-Mutafchieva L, Adams G, Edwards CJ, Moss J, Burger JA, et al. Mesenchymal precursor cells in the blood of normal individuals. Arthritis Research. 2000;**2**(6):477-488. DOI: 10.1186/ar130

[3] Erices A, Conget P, Minguell JJ. Mesenchymal progenitor cells in human umbilical cord blood. British Journal of Haematology. 2000;**109**(1):235-242

[4] Zuk PA, Zhu M, Ashjian P, De Ugarte DA, Huang JI, Mizuno H, et al. Human adipose tissue is a source of multipotent stem cells. Molecular Biology of the Cell. 2002;**13**(12):4279-4295

[5] McElreavey KD, Irvine AI, Ennis KT, McLean WH. Isolation, culture and characterisation of fibroblast-like cells derived from the wharton's jelly portion of human umbilical cord. Biochemical Society Transactions. 1991;**19**:29S

[6] De Bari C, Dell'Accio F, Tylzanowski P, Luyten FP. Multipotent mesenchymal stem cells from adult human synovial membrane. Arthritis and Rheumatism. 2000;**44**:1928-1942

[7] Pasquinelli G, Tazzari PL, Vaselli C, Foroni L, Buzzi M, Storci G, et al. Thoracic aortas from multiorgan donors are suitable for obtaining resident angiogenic mesenchymal stromal cells. Stem Cells. 2007;**25**:1627-1634

[8] Pasquinelli G, Pacilli A, Alviano F, Foroni L, Ricci F, Valente S, et al. Multidistrict human mesenchymal vascular cells: Pluripotency and stemness characteristics. Cytotherapy. 2010;**12**:275-287. DOI: 10.3109/14653241003596679

[9] Dominici M, Le Blanc K, Mueller I, et al. Minimal criteria for defining multipotent mesenchymal stromal cells. The international society for cellular therapy position statement. Cytotherapy. 2006;**8**:315-317

[10] Sidney LE, Branch MJ, Dunphy SE, Dua HS, Hopkinson A. Concise review: Evidence for CD34 as a common marker for diverse progenitors. Stem Cells. 2014;**32**(6):1380-1389

[11] Stzepourginski I, Nigro G, Jacob JM, et al. CD34+ mesenchymal cells are a major component of the intestinal stem cells niche at homeostasis and after injury. Proceedings of the National Academy of Sciences of the United States of America. 2017;**114**(4):E506-E513

[12] Eto H, Ishimine H, Kinoshita K, et al. Characterization of human adipose tissue-resident hematopoietic cell populations reveals a novel macrophage subpopulation with CD34 expression and mesenchymal multipotency. Stem Cells and Development. 2012;**22**(6):985-997

[13] Lin CS, Ning H, Lin G, Lue TF. Is CD34 truly a negative marker for mesenchymal stromal cells? Cytotherapy. 2012;**14**(10):1159-1163

[14] Wu Y, Chen L, Scott PG, Tredget EE. Mesenchymal stem cells enhance wound healing through differentiation and angiogenesis. Stem Cells. 2007;**25**(10):2648-2659. DOI: 10.1634/stemcells.2007-0226

[15] Xu L, Liu Y, Sun Y, et al. Tissue source determines the differentiation potentials of mesenchymal stem cells: A comparative study of human mesenchymal stem cells from bone

marrow and adipose tissue. Stem Cell Research & Therapy. 2017;**8**(1):275. DOI: 10.1186/s13287-017-0716-x

[16] Xu L, Zhou J, Liu J, et al. Different angiogenic potentials of mesenchymal stem cells derived from umbilical artery, umbilical vein, and Wharton's jelly. Stem Cells International. 2017;**2017**:3175748

[17] Prockop DJ, Oh JY. Mesenchymal stem/stromal cells (MSCs): Role as guardians of inflammation. Molecular Therapy. 2011;**20**(1):14-20

[18] Barry FP, Murphy JM. Mesenchymal stem cells: Clinical applications and biological characterization. The International Journal of Biochemistry & Cell Biology. 2004;**36**:568-584. DOI: 10.1016/j.biocel.2003.11.001

[19] Chamberlain G, Fox J, Ashton B, Middleton J. Concise review: Mesenchymal stem cells: Their phenotype, differentiation capacity, immunological features, and potential for homing. Stem Cells. 2007;**25**:2739-2749

[20] Lomakova YD, Londregan J, Maslanka J, Goldman N, Somerville J, Riggs JE. PHA eludes macrophage suppression to activate CD8+ T cells. Immunobiology. 2018. DOI: 10.1016/j.imbio.2018.10.004

[21] Selmani Z, Naji A, Zidi I, Favier B, Gaiffe E, Obert L, et al. Human leukocyte antigen-G5 secretion by human mesenchymal stem cells is required to suppress T lymphocyte and natural killer function and to induce CD4+CD25highFOXP3+ regulatory T cells. Stem Cells. 2008;**26**:212-222. DOI: 10.1634/stemcells.2007-0554

[22] Le Rond S, Azema C, Krawice-Radanne I, et al. Evidence to support the role of HLA-G5 in allograft acceptance through induction of immuno-suppressive/regulatory suppressive/regulatory T cells. Journal of Immunology. 2006;**176**:3266-3276

[23] Wang Y, Chen X, Cao W, Shi Y. Plasticity of mesenchymal stem cells in immunomodulation: Pathological and therapeutic implications. Nature Immunology. 2014;**15**(11):1009-1016. DOI: 10.1038/ni.3002

[24] Mattar P, Bieback K. Comparing the immunomodulatory properties of bone marrow, adipose tissue, and birth-associated tissue mesenchymal stromal cells. Frontiers in Immunology. 2015;**6**:560

[25] Li X, Bai J, Ji X, Li R, Xuan Y, Wang Y. Comprehensive characterization of four different populations of human mesenchymal stem cells as regards their immune properties, proliferation and differentiation. International Journal of Molecular Medicine. 2014;**34**:695-704. DOI: 10.3892/ijmm.2014.1821

[26] Bunnell BA, Betancourt AM, Sullivan DE. New concepts on the immune modulation mediated by mesenchymal stem cells. Stem Cell Research & Therapy. 2010;**1**:34. DOI: 10.1186/scrt34

[27] Waterman RS, Tomchuck SL, Henkle SL, Betancourt AM. A new mesenchymal stem cell (MSC) paradigm: Polarization into a pro-inflammatory MSC1 or an Immunosuppressive MSC2 phenotype. PLoS One. 2010;**5**:e10088

[28] Bernardo ME, Fibbe WE. Mesenchymal Stromal Cells: Sensors and Switchers of Inflammation. Cell Stem Cell. 2013;**13**(4):392-402. DOI: 10.1016/j.stem.2013.09.006

[29] Li W, Ren G, Huang Y, Su J, Han Y, Li J, et al. Mesenchymal stem cells: A double-edged sword in regulating immune responses. Cell Death and Differentiation. 2012;**19**:1505. DOI: 10.1038/cdd.2012.26

[30] Djouad F, Fritz V, Apparailly F, Louis-Plence P, Bony C, Sany J, et al. Reversal of the immunosuppressive properties of mesenchymal stem cells by tumor necrosis factor α in collagen-induced arthritis. Arthritis and Rheumatism. 2005;**52**:1595-1603. DOI: 10.1002/art.21012

[31] Li B, Fu J, Chen P, Zhuang W. Impairment in immunomodulatory function of mesenchymal stem cells from multiple myeloma patients. Archives of Medical Research. 2010;**41**(8):623-633. DOI: 10.1016/j.arcmed.2010.11.008

[32] Griffin MD, Elliman SJ, Cahill E, English K, Ceredig R, Ritter T. Concise review: Adult mesenchymal stromal cell therapy for inflammatory diseases: How well are we joining the dots? Stem Cells. 2013;**31**:2033-2041. DOI: 10.1002/stem.1452

[33] Singer NG, Caplan AI. Mesenchymal Stem Cells: Mechanisms of Inflammation. Annual Review of Pathology: Mechanisms of Disease. 2011;**6**(1):457-478

[34] Valente S, Alviano F, Ciavarella C, Buzzi M, Ricci F, Tazzari PL, et al. Human cadaver multipotent stromal/stem cells isolated from arteries stored in liquid nitrogen for 5 years. Stem Cell Research & Therapy. 2014;**5**(1):8

[35] Valente S, Ciavarella C, Pasanisi E, Ricci F, Stella A, Pasquinelli G. Hepatocyte growth factor effects on mesenchymal stem cells derived from human arteries: A novel strategy to accelerate vascular ulcer wound healing. Stem Cells International. 2016;**2016**:3232859. DOI: 10.1155/2016/3232859

[36] Ciavarella C, Alviano F, Gallitto E, Ricci F, Buzzi M, Velati C, et al. Human vascular wall mesenchymal stromal cells contribute to abdominal aortic aneurysm pathogenesis through an impaired immunomodulatory activity and increased levels of matrix metalloproteinase-9. Circulation Journal. 2015;**79**(7):1460-1469

[37] Ciavarella C, Gallitto E, Ricci F, Buzzi M, Stella A, Pasquinelli G. The crosstalk between vascular MSCs and inflammatory mediators determines the pro-calcific remodelling of human atherosclerotic aneurysm. Stem Cell Research & Therapy. 2017;**8**(1):99. DOI: 10.1186/s13287-017-0554-x

[38] Tang Z, Wang A, Yuan F, et al. Differentiation of multipotent vascular stem cells contributes to vascular diseases. Nature Communications. 2012;**3**:875. DOI: 10.1038/ncomms1867

[39] Kramann R, Goettsch C, Wongboonsin J, et al. Adventitial MSC-like cells are progenitors of vascular smooth muscle cells and drive vascular calcification in chronic kidney disease. Cell Stem Cell. 2016;**19**(5):628-642

[40] Leszczynska A, O'Doherty A, Farrell E, Pindjakova J, O'Brien FJ, O'Brien T, et al. Differentiation of vascular stem cells contributes to ectopic calcification of atherosclerotic plaque. Stem Cells. 2016;**34**:913-923. DOI: 10.1002/stem.2315

[41] Kroesen BJ, Teteloshvili N, Smigielska-Czepiel K, et al. Immuno-miRs: Critical regulators of T-cell development, function and ageing. Immunology. 2014;**144**(1):1-10

[42] Matysiak M, Fortak-Michalska M, Szymanska B, Orlowski W, Jurewicz A, Selmaj K. MicroRNA-146a negatively regulates the immunoregulatory activity of bone marrow stem cells by targeting prostaglandin E synthase-2. Journal of Immunology. 2013;**190**(10):5102

[43] Xu C, Ren G, Cao G, et al. miR-155 regulates immune modulatory

properties of mesenchymal stem cells by targeting TAK1-binding protein 2. The Journal of Biological Chemistry. 2013;**288**(16):11074-11079

[44] Abdi J, Rashedi I, Keating A. Concise review: TLR Pathway-miRNA interplay in mesenchymal stromal cells: Regulatory roles and therapeutic directions. Stem Cells. 2018;**36**:1655-1662. DOI: 10.1002/stem.2902